STUDY GUIDE

Richard M. Busch

Seventh Edition

Earth

AN INTRODUCTION TO PHYSICAL GEOLOGY

Tarbuck & Lutgens

Prentice
Hall

PRENTICE HALL, Upper Saddle River, NJ 07458

Senior Editor: Patrick Lynch
Associate Editor: Amanda Griffith
Assistant Managing Editor: Dinah Thong
Executive Managing Editor: Kathleen Schiaparelli
Production Editor: Patty Donovon/Pine Tree Composition
Supplement Cover Manager: Paul Gourhan
Manufacturing Buyer: Lynda Castillo
Cover Photo: Galen Rowell/Mountain Light Photography
Illustrations: Dennis Tasa

© 2002, 1999, 1996 by Prentice-Hall, Inc.
Upper Saddle River, NJ 07458

Printed in the United States of America

10 9 8 7 6 5 4 3 2

ISBN 0-13-092031-2

Pearson Education Ltd., *London*
Pearson Education Australia Pty. Limited, *Sydney*
Pearson Education Singapore, Pte. Ltd.
Pearson Education North Asia Ltd. *Hong Kong*
Pearson Education Canada, Ltd., *Toronto*
Pearson Educatión de Mexico, S. A. de C. V.
Pearson Education—Japan, *Tokyo*
Pearson Education Malaysia, Pte. Ltd.
Pearson Education, *Upper Saddle River, New Jersey*

CONTENTS

STUDENT INSTRUCTIONS

HOW TO STUDY AND USE THIS GUIDE EFFECTIVELY

Students commonly ask several questions about a course stressing physical geology: *How can I get an A in the course? What is the best way to learn geology?* and *How should I study?* There is no perfect answer to any of these questions. However, successful students use many of the same study techniques and have many study habits in common.

These instructions point out some of the study techniques and habits that you can use and develop to help you learn more about physical geology in conjunction with the textbook *Earth: An Introduction to Physical Geology,* Seventh Edition, by E. J. Tarbuck and F. K. Lutgens.

WHICH WAY TO AN A?

The best way to attain an A in any course is by thoroughly understanding the topics addressed in the course. Therefore, each chapter of this study guide starts with a set of learning objectives—the topics you should know and understand. The learning objectives are followed by a vocabulary review to help you determine how well or poorly you understand the most basic elements of the chapter material. You should also be able to apply what you have learned to solve problems and make sound judgments. Therefore, each chapter of the study guide contains activities and problems relative to the material you should have learned. Finally, a review exam is provided for you to determine the extent of your total understanding of a chapter. This will help you determine how prepared you are for an actual exam on the chapter.

HOW TO LEARN PHYSICAL GEOLOGY

Learning is the process of gaining knowledge, understanding, and abilities related to a topic through study, instruction, and experience. As a student, you should emphasize all three of these aspects of the learning process as much as possible. This means that you should study your book, carefully consider classroom discussions, take careful notes, and carefully observe, handle, and/or examine any physical or visual materials presented by your instructor.

Learning physical geology also means that you need to develop and use certain habits and process skills that characterize scientific thinking and learning:

1. **Think of yourself as a scientist!** By the end of the course you will be a geologist.

2. **Be prepared for class!** Study in a location that is quiet and organized. Clean off your desk, adjust the lighting to a comfortable level, and concentrate on reading and understanding the textbook, CD-ROM, or Web-based materials that were assigned for class. Work at your own comfortable pace.

3. **Be curious and proactive!** Ask your instructor questions about key terms, concepts, objects, and ideas that you do not understand. If objects or visual aids are provided in class or laboratory, then

pay particular attention to how those materials look, feel, sound, smell, or even taste. If a concept or observation does not make sense to you, then ask the instructor to explain or to present it differently. Study the CD-ROM materials related to each topic, even if your instructor did not assign them. You have paid time and money to be in the class, so use your time wisely and productively. The instructor is there to help you, but you must seek his or her help! Complete assignments on time, and attend all classes.

4. **Make connections to your own life!** Structure your learning by linking new ideas and experiences to what you already know, understand, and are able to do. Think of ways that topics affect you and your community. Think of ways to use what you learn. For example, consider how geology topics relate to your major, careers, prices, public policy decisions and laws, public safety, quality of life, and the future of Earth.

5. **Think of science as a sport!** Be persistent, and try hard to win new ideas, new abilities, and good grades! Study hard and repeat questions, examples, or experiments until you obtain meaningful answers to your questions and understand ideas or observations in your own mind. Winning good grades is not easy. No pain, no gain!

6. **Make careful observations and classify things!** Note properties of objects (look, feel, sound, smell, taste), how things change through time, rhythms or cycles of change, similarities, and differences. Group objects or events on the basis of their similarities and differences, and arrange events relative to their time and place of occurrence.

7. **Quantify relationships!** Review common systems and units of measurement (such as the metric system), plus common tools of measurement (such as meter sticks, yardsticks, protractors, gram balances, etc.). Note and compare quantities and dimensions of objects. Note measurable changes through time.

8. **Communicate effectively!** Put definitions, ideas, and information in sentences, charts, graphs, maps, or sketches that help you understand them most easily. Keep your class handouts in order relative to your notes.

9. **Think critically and make inferences!** Consider all of the evidence or information available before you answer a question or form a conclusion, Change or modify your answers/conclusions as necessary on the basis of new information and/or evidence. Make predictions based on what you learn.

10. **Use this study guide effectively!** Read on to find out how.

HOW TO USE THIS STUDY GUIDE EFFECTIVELY

For each chapter of your textbook there is a corresponding chapter (of the same number) in this study guide. For example, chapter 1 of the study guide relates to chapter 1 of your textbook, and so on.

Each chapter of this study guide is composed of five parts. The meaning and effective use of these five parts is discussed next.

LEARNING OBJECTIVES

These objectives describe what you should know, understand, and be able to do after you have read and studied a textbook chapter or CD-ROM materials and completed the corresponding chapter in this study guide.

You should read the list of learning objectives for a chapter before you actually read that textbook chapter and again after you have read it. If you answer the review questions at the end of a textbook chapter, then you should consider your answers relative to the learning objectives.
You should also review each term in the list of key terms that is located at the end of each textbook chapter. As you review the terms, consider how each one relates to the learning objectives. Remember, it is of little use to memorize terms and definitions if you do not have a reason for doing so (as given in the Learning Objectives).

VOCABULARY REVIEW

The Vocabulary Review is composed of fill-in-the-blank statements that are designed to test your basic knowledge of a chapter (mainly key terms). If you have trouble with this part, then you should again review the key terms listed at the end of the chapter. You may even choose to prepare and use flash cards, having a key term on the front side and the term's definition on the reverse side. Study the terms by quizzing yourself, using the cards. That is, read a word, then recall its definition before checking on the back of the card. Next, read the definition, then recall the word before checking on the front of the card. When you feel that you know the flash cards forward and backward, then try the Vocabulary Review again. If you do well, then move on to Applying What You Have Learned. If you do not do well, then study your flash cards more.

APPLYING WHAT YOU HAVE LEARNED

For this part of each study guide chapter, you will be asked to label diagrams, fill in charts, or fill in the blanks. All of these tasks are designed to test your comprehension of key terms and other basic knowledge. If you have trouble with this part, then you do not fully understand the key terms or basic knowledge of that chapter in your own words or in symbolic form (pictures, charts). If this is the case, then you may choose to prepare another set of flash cards. This time write a key term or idea on the front of each card. Then, on the reverse side of each card, write a definition in your own words and/or make a sketch (or rephrase the idea in your own words or in a sketch). Perform this activity until you feel that you actually understand the key terms and ideas in your own mind and in written and symbolic form, then try Applying What You Have Learned again. If you do well, then move on to the Activities and Problems. If you do not do well, then revise your flash cards and study them more.

ACTIVITIES AND PROBLEMS

The Activities and Problems section of each study guide chapter is designed to find out how well analyze problems (i.e., break problems up into component parts), find answers to problems (i.e., put together the component parts to make an entire idea or judgment), and make predictions (i.e., apply your knowledge and understanding to a different setting or a future situation that might occur and present it persuasively). If you have trouble with this part, then you should restudy all examples in the textbook chapter that dealt with problem-solving or practical applications relating the key terms to the learning objectives. You may also choose to prepare still another set of flash cards for the chapter. Write a key term or idea on the front of each, card. Then, on the reverse side of each card, write down how the term or idea relates to the learning objectives. Perform this activity, and even revise and restudy your flash cards as necessary, until you do well on the Activities and Problems.

REVIEW EXAM

The Review Exam of each study guide chapter is designed to test your total knowledge and understanding of a textbook chapter. It is one way for you to determine how ready you are to take an actual exam on the chapter. If you do poorly on the review exam, then you should again review your notes and flash cards for that chapter. You may also consult with your instructor for more tips about studying and learning what was presented in that chapter of your textbook or in class.

ANSWER KEY

Answers for each question of the Vocabulary Reviews, Applying What You Have Learned, Activities and Problems, and Review Exams are provided in the Answer Key at the end of the study guide. The answers are arranged by their chapter, section, and number.

To find a correct answer, refer to the example below of a fill-in-the-blank statement from the Vocabulary Review section of Chapter 1.

1. We live on the planet called (a)_____.

↑Blank space for your answer, in pencil.

The correct answer to this item can be found in the Answer Key at the rear of the study guide. To find the answer, go to the answers for Chapter 1, item 1. a, in the back of the book.

WARNING

If you have not read all of the student instructions,
then you should do so now, before you proceed.
The student instructions start on page iv.

CHAPTER 1: AN INTRODUCTION TO GEOLOGY

LEARNING OBJECTIVES

After reading and studying this chapter, you should be able to

1.   Distinguish between physical and historical geology.

2.   Contrast catastrophism and uniformitarianism.

3.   Describe James Hutton's contributions to modern geology.

4.   Describe how nineteenth-century scientists developed a geologic time scale using principles of relative dating, such as the law of superposition and the principle of fossil succession.

5.   List the basic elements of the scientific method and how it is related to development of a hypothesis (or model), theory, and paradigm.

6.   List and describe the four main spheres of the Earth system: solid Earth, hydrosphere, atmosphere, and biosphere.

7.   Understand the meaning of the statement "Earth is a system."

8.   Name and describe the two main sources of energy that power the Earth system and drive its external and internal processes.

9.   Know the major continental features (mountains, shields) and seafloor features (continental shelf, continental slope, trenches, oceanic ridge system).

10.   Outline the nebular hypothesis.

11.   Name and describe the compositional and structural subdivisions of Earth's interior.

12.   Describe the theory of plate tectonics.

13.   List and distinguish among three basic types of plate boundaries.

14.   Sketch and label a diagram of the rock cycle and understand how the rock cycle is related to Earth as a system.

15.   Outline how the rock cycle relates to the plate tectonics theory at a convergent plate boundary along a continental margin.

16.   Define and understand the key terms listed at the end of the chapter.

Vocabulary Review

1.   The Earth's physical environment is traditionally divided into three major parts. In addition to the solid portion of the Earth (called the lithosphere), the water portion of the planet is called the (a)_____, and the envelope of air around the planet is called the (b)_____.

2. The compositional divisions of the solid portion of the Earth have specific names. The solid iron-rich shell at Earth's center is the (a)_____. The molten metallic shell is the (b)_____. The solid rocky division overlying the molten metallic shell is the (c)_____. The least dense, outer shell on which people live is the (d)_____.

3. Geology is a word that means _____.

4. The science of geology is traditionally divided into two broad areas.
(a)_____ geology examines the materials composing the Earth and seeks to understand the processes that operate on and beneath its surface.
(b)_____ geology seeks to understand the origin of the earth and its development through time.

5. The doctrine of _____ is that the Earth's landscape has been shaped primarily by sudden and often worldwide disasters of unknown causes.

6. The doctrine of uniformitarianism states that (a)_____
_____.
This doctrine was put forth by (b)_____.

7. A _____ is a group of interacting or independent parts that form a complex whole.

8. James Ussher, Anglican Archbishop of Armagh, determined that the Earth was created in (a)_____ B.C.; however, current scientific estimates put the age of the Earth at about (b)_____ years.

9. (a)_____ dating means placing events in their proper sequence, or order, without knowing their age in years. This is done by applying principles such as the law of superposition, which states that (b)_____
_____
_____ Furthermore, the principle of (c)_____ states that fossil organisms succeed one another in a definite and determinable order.

10. A nebula is a_____
_____.

11. Scientists try to explain how or why things happen in the manner observed. They do this by constructing a tentative (or untested) explanation, called a(n) (a)_____.
If such a tentative explanation survives extensive testing and scrutiny, and competing explanations are eliminated, then the explanation is called a scientific (b)_____.

12. The _____ suggests that all bodies of the solar system formed from an enormous cloud composed mostly of hydrogen and helium.

13. The soft, weak structural layer ("weak zone") of the Earth's mantle that is actually capable of flow is called the (a)_____. The more rigid layer beneath it, where rocks are very hot and capable of very gradual flow, is the (b) _____. The cool and very rigid sphere of rock situated just above the weak zone is called the (c)_____.

14. The most prominent mountain range on the Earth is the (a)_____ By contrast, narrow grooves, in places more than 11,000 meters deep, occur in the sea floor adjacent to volcanic islands and to young mountains on the edges of continents. These deep, narrow grooves are called (b)_____.

15. According to the theory of (a)_____, the rigid lithosphere of the Earth is broken into numerous segments called (b)_____. The three distinct types of boundaries between these segments of rigid lithosphere are (c)_____ boundaries, (d)_____ boundaries, and (e)_____ boundaries.

16. As new oceanic lithosphere is created at the ocean ridges, the old oceanic lithosphere is pushed aside. This process is called (a)_____ and occurs at a rate of (b)_____ per year.

17. Stable continental interiors are known as _____.

18. Regions where oceanic lithosphere is being consumed and melted are called _____.

19. Molten rock material beneath the Earth's surface is called (a)_____. When it cools, it solidifies by the process of crystallization and forms (b)_____ rock.

20. If rocks are exposed at the surface of the Earth, then they slowly decompose and disintegrate by the process of (a)_____, which produces particles called (b)_____. These particles may be converted into rock if they undergo (c)_____, and so become (d)_____ rock.

21. If rocks react to great pressures and heat, then they may change into _____ rocks.

22. Along most coasts a gently sloping platform extends seaward from the shore and is called the (a)_____. At the seaward edge of such platforms, a steep drop-off, called a (b)_____, occurs.

23. Theories that become extensively documented with a high degree of confidence are called
_____, because they explain a large number of interrelated aspects
of the natural world.

24. The Earth system has four main interacting spheres. Earth's water is its
(a) _____. Earth's air is its (b) _____.
Earth's life is its (c) _____. The largest sphere lies beneath all of the
others and is called the (d) _____.

Applying What You Have Learned

1. Earth is a system. What does this mean? _____
_____
_____

2. List the four basic steps of the scientific method. _____
_____
_____
_____

3. On the seafloor profile below, fill in the blanks with the correct name of the feature that is-
labeled.

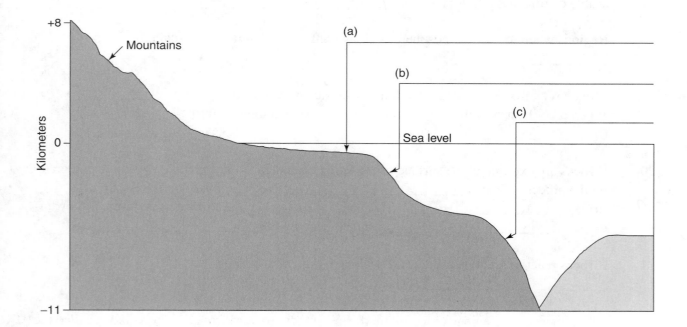

4. Fill in the blanks with the correct name of the feature that is labeled.

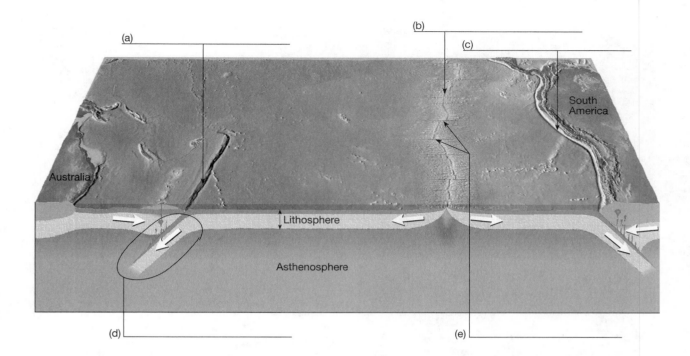

(a) _____

(b) _____

(c) _____

South America

Lithosphere

Asthenosphere

Australia

(d) _____

(e) _____

5. Describe what is happening in each stage of the nebular hypothesis pictured below.

(a)

(b)

(c)

(d)

6. In the spaces provided, describe what happens during the rock cycle.

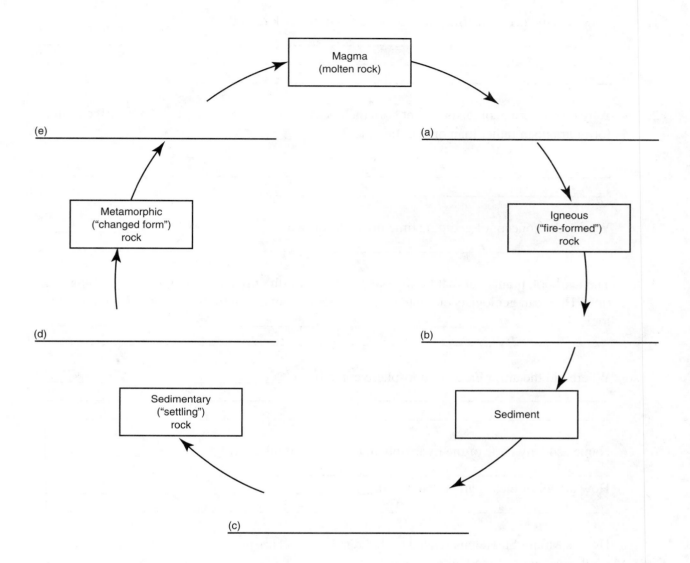

ACTIVITIES AND PROBLEMS

1. How was the geologic time chart derived from the rock record? _____
    _____
    _____
    _____

2. Why do modern geologists accept both the doctrine of catastrophism and the doctrine of uniformitarianism rather than one or the other?_____
    _____
    _____
    _____

3. Why is the concept of geologic time difficult for people to conceive of? _____
    _____

4. The textbook points out that humans appeared on Earth only recently in terms of geologic time. How can geologists claim that there were no humans on Earth during much older geologic times? _____
    _____

5. Where did the air in Earth's atmosphere come from? _____
    _____
    _____

6. Name as many parts of the hydrosphere as you can think of. (a)_____
    _____
    How are all of these parts related? (b)_____
    _____

7. How is Mount St. Helens related to the trench that occurs just west of it, close to the northwest coast of the United States?_____
    _____
    _____

8. Recall the rock cycle (refer back to textbook, if necessary) and the idea that sediments are derived by weathering of igneous rocks. Is this the only way sediments form? Explan_____
    _____
    _____
    _____

9. All science is based on one assumption. What is it? _____
    _____

10.   The Earth system is powered by energy from two sources. What are they?

_____

_____

REVIEW EXAM

1.   Why are there so few craters on the surface of the Earth? _____

_____

_____

2.   The lowest features on the surface of Earth are _____

3.   Circle the letter of the most correct answer. The lithosphere is

      a.   the outermost skin of the solid Earth.
      b.   less than 100 km thick.
      c.   moved by convection currents in the asthenosphere.
      d.   all of the above.
      e.   a and b.

4.   Circle the letter of the most correct answer. The oceanic ridge system is

      a.   a zone where crustal plates diverge.
      b.   a zone where crustal plates merge and subduct.
      c.   a zone of constructional processes.
      d.   a and c.
      e.   b and c.

5.   Circle the letter of the most correct answer. Catastrophism is the idea that

      a.   Earth's landscapes have been shaped by sudden worldwide disasters produced by un-known causes that no longer operate.
      b.   physical, chemical, and biological laws that operate today have also operated in the geologic past.
      c.   Earth's landscapes have been shaped by the influence of the stars.
      d.   catastrophies will happen more often each century until Earth experiences a world-wide disaster.

6. Circle the letter of the most correct answer. Uniformitarianism is the idea that

   a. Earth's landscapes have been shaped by sudden worldwide disasters produced by unknown causes that no longer operate.
   b. physical, chemical, and biological laws that operate today have also operated in the geologic past.
   c. Earth's landscapes have been shaped by the influence of the stars.
   d. catastrophies will happen more often each century until Earth experiences a worldwide disaster.

7. Circle the letter of the most correct answer. The person who effectively demonstrated that geological processes occur over extremely long periods of time was

   a. Hutton.
   b. Aristotle.
   c. Ussher.

8. Circle the letter of the most correct answer. The San Andreas fault is what kind of plate boundary?

   a. convergent
   b. transform
   c. divergent

9. Circle the letter of the most correct answer. Seafloor spreading begins at

   a. divergent plate boundaries.
   b. trenches.
   c. transform fault boundaries.
   d. the inner core.

10. For each of the following statements, write T for *"true"* or F for *"false"* on the space provided to indicate its validity.

    (a) _____ The densest portion of the Earth is its mantle.
    (b) _____ All science is based on the assumption that the natural world behaves in a consistent and predictable manner.
    (c) _____ Shields and sea floors are the oldest features of the Earth's surface.
    (d) _____ Changes in weather and climate significantly affect how the Earth's surface is shaped.
    (e) _____ The density of continental rocks is generally less than the density of seafloor rocks.
    (f) _____ The positions of shorelines have not varied much throughout most of geologic time.
    (g) _____ The San Andreas fault of California is a transform boundary.

(h) _____ When continents collide, very rapid subduction occurs.
(i) _____ Erosion and volcanism are destructional processes.
(j) _____ The primary tool used to subdivide geologic time is fossils.

11. Fill in the blanks with correct names of the structural and compositional layers of the Earth.

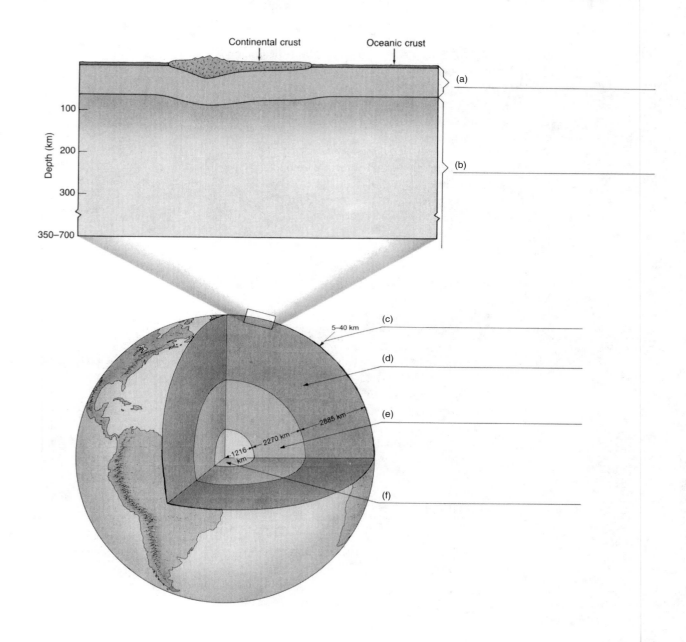

CHAPTER 2: MATTER AND MINERALS

LEARNING OBJECTIVES

After reading and studying this chapter, you should be able to

1. Distinguish between the terms *minerals* and *rocks*.

2. List the three main particles of an atom and explain how they differ from one another.

3. Contrast ionic, covalent, and metallic bonding.

4. List and define the physical properties that are useful in identifying minerals.

5. List (in decreasing abundance) the eight most common elements of the continental crust.

6. Briefly describe the silicon-oxygen tetrahedron.

7. List and distinguish among the common silicate minerals.

8. Identify minerals in textbook Table 2.4 that are used in your home and explain how you use them.

9. Define and understand the key terms listed at the end of the chapter.

10. Understand what gemstones are and what qualities make gemstones valuable.

11. Be able to comment on the health risks associated with the use of asbestos.

VOCABULARY REVIEW

1. The study of minerals is called _____.

2. Minerals are made up of pure substances called (a)_____ and aggregates of minerals are called (b)_____.

3. Two minerals having the same chemical composition but different crystalline structures are called _____.

4. The external shape of a mineral crystal, which reflects its orderly internal arrangement of atoms, is called the mineral's _____.

5. The smallest part of an element (that still retains the element's properties) is called a(n) _____.

6. The way light is reflected from the surface of a mineral is called _____

7. The color of a mineral in powdered form is called _____.

:gion, called the (a)_____, which contains
:s called (b)_____ and neutral particles
_____. Negatively charged particles orbiting the nucleus
_____.

the nucleus of an atom determines the (a)_____
_____ is determined by adding together the
utrons.

it having different numbers of neutrons in the nucleus are called
___.

int elements in the continental crust of the Earth in decreasing

_____

_____

_____

:harge is called a(n) (a)_____ and the type of
he electronic attraction of oppositely charged atoms is called
____ bonding.

.l to abrasion or scratching is called (a)_____,
: to (b)_____scale.

planes of weak bonding (flat surfaces) are said to have a mineral
_____. Minerals that do not break along planes of
said to (b)_____ when they break.

ince composed of two or more elements is called _____
_____.

a mineral to the weight of an equal volume of water is what
_____.

a process of natural disintegration called
_____.

d at given distances from the nucleus in regions called
_. To be chemically stable like the noble gases, every atom
how many electrons? (b)_____.

ling process are called _____ electrons.

20. The type of bonding that forms when atoms share electrons to acquire the stable noble gas arrangement is called (a)_____ bonding. But if the electrons move freely from one atom to another, then there is said to be (b)_____ bonding.

21. Dark colored minerals containing iron and/or magnesium are called _____ minerals.

22. Any solid mass of mineral or mineral-like matter that occurs naturally is called a(n) _____.

23. The most common mineral group in the crust of the Earth is the (a)_____. All of them have the same atomic building block, which is called the (b)_____.

24. Name the type of silicate structure that is present in each mineral below:

    quartz   (a) _____
    mica   (b) _____
    feldspar   (c) _____
    olivine   (d) _____
    pyroxene   (e) _____
    amphibole   (f) _____

25. Name the mineral that is used to make the items listed below:

    plaster of Paris   (a) _____
    common table salt   (b) _____
    pencil lead   (c) _____
    aluminum foil   (d) _____
    baby powder   (e) _____

APPLYING WHAT YOU HAVE LEARNED

1. Examine the diagram of an atom below. Its nucleus contains 8 protons and 8 neutrons.

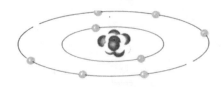

Is this atom an ion? Explain. (a) _____

_____

_____

How many valence electrons does this atom need to satisfy the octet rule?   (b) _____

How many bonds will this atom form?   (c) _____

What is the mass number of this atom?   (d) _____

What is the atomic number of this atom?   (e) _____

How many energy-level shells does this atom have?   (f) _____

2.   Examine the diagram of an atom below. Its nucleus contains 17 protons.

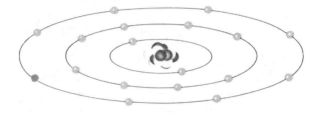

Is this atom an ion? Explain. (a) _____

_____

_____

How many valence electrons does this atom need to satisfy the octet rule? (b) _____

How many energy-level shells does this atom have? (c) _____

3.   For each illustration below, note the number of cleavage directions.

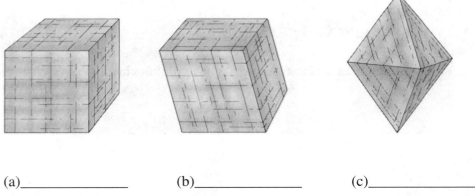

(a)_____   (b)_____   (c)_____

4.   What is a mineral?_____

_____

_____

5. Imagine a piece of broken glass, as at the end of a broken glass bottle. How would you describe the way glass breaks, specifically in terms of the broken surfaces? _____
_____

6. What is cleavage? _____
_____
_____

ACTIVITIES AND PROBLEMS

1. Is ice a mineral? Explain your answer. _____
_____
_____

2. Is sugar a mineral? Explain your answer. _____
_____
_____

3. How can you tell ferromagnesian silicates from nonferromagnesian silicates? _____
_____
_____

4. If an atom of carbon has 8 neutrons, 6 electrons, and 6 protons, then what is the atom's atomic mass? (a)_____ If another atom has 6 neutrons, 6 electrons, and 6 protons, then what is that atom's atomic mass? (b)_____ Are both of these atoms carbon? (c)_____ Why or why not? (d)_____
_____
_____

5. If a 2-liter bottle of water weighs 2 kilograms, and a piece of native copper weighing 17.8 kilograms has a volume of 2 liters, then what is the specific gravity of copper? _____

6. Imagine that a crystal of plagioclase feldspar is forming inside a cooling magma body and that it has a chemical formula of $CaAl_2Si_2O_8$. If all of the available Ca atoms are used up, then what element is likely to substitute for the Ca as the plagioclase feldspar crystal continues to grow? (a)_____ Why? (b)_____
_____

If such a substitution actually occurred, then what else might change about the new kind of feldspar so formed? (c)_____

7. Why does blue and brown asbestos pose a significant health risk, while the white Chrysotile asbestos does not? _____

_____

REVIEW EXAM

1. The element that commonly substitutes for Fe is _____.

2. Sheet silicates generally have how many prominent cleavage directions? _____

3. A silicate mineral having two cleavage directions and striations on some of the cleavage surfaces is the mineral _____.

4. A green-colored, single-tetrahedron silicate mineral that forms at high temperatures is

_____.

5. A silicate mineral having two prominent cleavage directions that intersect at angles of 60 degrees and 120 degrees is _____.

6. Other than beauty, what aspect of gemstones adds most to their value?

_____

7. Amethyst is a purple gemstone variety of what mineral? _____

8. What is the difference between a karat and a carat? _____

9. Unstable isotopes disintegrate through a process called _____.

10. Is muscovite a ferromagnesian silicate mineral or a nonferromagnesian mineral? (a) _____
Explain your answer. (b)_____

_____

11. What are polymorphs? _____

_____

12. What is the most abundant mineral in the crust of the earth (making up more than 50% of the crust by weight)? _____

13. What minerals typically form from chemical weathering of feldspars and muscovite?

_____

14. The only silicate mineral made entirely of silicon and oxygen is _____.

15. Circle the letter of the most correct answer. All silicate mine

    a.   They all contain Al.
    b.   They all contain silicon-oxygen tetrahedra.
    c.   They all contain iron and magnesium.
    d.   Answers a and b are both correct.
    e.   All of these answers are correct.

16. Circle the letter of the most correct answer. Pyroxene is

    a.   a ferromagnesian silicate.
    b.   an ore of iron.
    c.   an ore of lead.
    d.   a and b.
    e.   all of the above.

17. Circle the letter of the most correct answer. The two most ab nental crust are

    a.   nitrogen (N) and oxygen (O).
    b.   oxygen (O) and silicon (Si).
    c.   silicon (Si) and iron (Fe).
    d.   silicon (Si) and Aluminum (Al).

18. Circle the letter of the most correct answer. Which mineral t

    a.   muscovite mica
    b.   feldspar
    c.   quartz
    d.   halite (table salt)

19. For each of the following statements, write T for *True* or F fo indicate its validity.

    (a) _____ Quartz is harder than feldspar.
    (b) _____ Talc is harder than calcite.
    (c) _____ Isotopes of the same element have equal number bers of protons.
    (d) _____ If an atom has 15 electrons, then it seeks three el
    (e) _____ Luster is the color of a mineral in powdered for
    (f) _____ Minerals with crystal form always have good cle
    (g) _____ Muscovite and biotite have the same physical pr different.

20. Fill in the table below on silicate minerals.

Silicate structure	Oxygen to silicon ratio	mineral	cleavage
oxygen atoms / silicon atom	4:1	olivine	(a)
	(b)	(c)	two planes at right angles
	(d)	(e)	(f)

20

CHAPTER 3: IGNEOUS ROCKS

LEARNING OBJECTIVES

After reading and studying this chapter, you should be able to

1. Know the difference between the terms *magma* and *lava*.

2. Describe how the rate of cooling and mobility of ions influence the sizes of crystals in igneous rocks.

3. List five different igneous rock textures and explain their origins.

4. Discuss the contributions of N. L. Bowen to our understanding of igneous rocks.

5. Name and describe the processes by which a single magma may produce several different igneous rock types.

6. Contrast the granitic (felsic), andesitic (intermediate), and basaltic (mafic) compositional groups and give the names of an aphanitic (fine-grained) and phaneritic (coarse-grained) representative for each group.

7. Describe three ways that magma can be generated from solid rock of the mantle, or lower crust.

8. Define and understand the key terms listed at the end of the chapter.

VOCABULARY REVIEW

1. Magma is a hot fluid (molten rock) that may contain what two things in addition to liquid?
   _____

2. Rocks that form when magma solidifies deep beneath the surface of the land are called
   (a)_____ or (b)_____ igneous rocks.

3. When lava solidifies (at or on the surface of the land), the rocks are classified as
   (a)_____ or (b)_____ igneous rocks.

4. The process by which ions arrange themselves into orderly patterns is called
   _____.

5. If lava is quenched, and ions of the lava have not had enough time to arrange themselves into an orderly crystalline structure, then a(n) (a)_____ texture results. The name for an igneous rock having such texture is (b)_____.

6. A coarse-grained igneous rock is said to have a(n) (a)_____ texture, whereas a fine-grained igneous rock is said to have a(n) (b)_____texture.

7. The process by which early-formed minerals settle to the bottom of the magma chamber in which they formed is called _____.

8. Voids (holes) left when gases escape from a lava flow are called
(a)_____. A rock containing such holes is said to have a(n)
(b) _____ texture.

9. A decrease in pressure is one factor that can cause rock of the mantle or lower crust to melt and form magma. When this occurs, it is called _____.

10. An igneous rock composed of two distinctly different sizes of mineral crystals is called
(a)_____ and is said to have a(n) (b)_____
texture. In such a rock, the large mineral crystals are called (c)_____ and
the small crystals forming the matrix are called the (d) _____

11. Igneous rocks composed of rock fragments ejected from volcanoes are said to have
a(n)_____ texture.

12. As magma cools, those minerals with higher melting points crystallize before minerals with lower melting points. The early-formed mineral crystals react with remaining magma at lower temperatures to form a series of different minerals. This is called

_____.

13. A coarse-grained igneous rock composed mostly of feldspar and quartz is called
(a)_____, but its fine-grained equivalent is called (b)_____.

14. Very dark-colored fine-grained igneous rocks are called _____.

15. The igneous rock called _____ will often float when placed in water.

16. The process of developing more than one rock type from a common magma is called

_____.

17. Exceptionally coarse-grained, crystalline igneous rocks have a(n) _____
texture.

18. The process believed to occur when one ascending magma body overtakes, and interacts with, another magma body is called _____.

19. The gradual increase in temperature with depth in the Earth is known as the_____

_____.

20. As rocks having mixtures of minerals are heated, some minerals melt at low temperatures, whereas others melt at higher temperatures. This is known as the process of

_____.

APPLYING WHAT YOU HAVE LEARNED

1. On the basis of mineral composition, fill in the boxes provided at the right hand side of the diagram below with the correct name of an igneous rock.

Bowen's Reaction Series

	Name of intrusive rock type	Name of extrusive rock type
(a)		komatiite
(b)		(c)
(d)		(e)
(f)		(g)

ACTIVITIES AND PROBLEMS

1. What is the difference between magma and lava?_____
   _____
   _____
   _____

2. How does rate of cooling affect the sizes of crystals in igneous rocks?_____
   _____
   _____
   _____

3. How does the viscosity of lava affect its crystallization?_____
   _____
   _____
   _____
   _____

4.   To the right of each picture of an igneous rock sample below, *name* the rock's texture and describe its *origin* (how the texture formed). All rocks are shown at their actual sizes.

SAMPLE   TEXTURE NAME AND ORIGIN

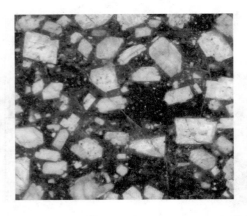

(a)_____

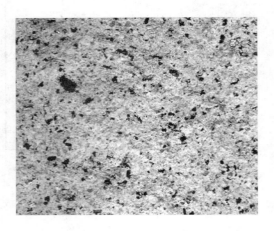

(b)_____

(c)_____

SAMPLE   TEXTURE NAME AND ORIGIN

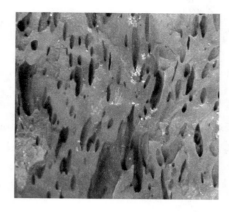

                                            (d)_____

5.   Reexamine rock sample 4(a) pictured on the previous page. Based on its proportion of mafic
     and felsic mineral crystals, this rock belongs to the (a) _____
     compositional group. The specific name of this rock (that describes its composition and grain
     size) is (b) _____.

6.   What are the three main discoveries that N. L. Bowen made about minerals crystallizing from
     a magma/lava? _____
     _____
     _____
     _____

7.   Describe three ways that magma can be generated from solid rock of the mantle and lower
     crust. _____
     _____
     _____
     _____

REVIEW EXAM

1.   Circle the letter of the most correct answer. A heated crystalline solid melts when

     a.   the ions vibrate so much that the vibrating breaks some of the chemical bonds.
     b.   the chemical bonds melt.
     c.   heat expansion of the chemical bonds causes them to break.
     d.   heat expansion of the ions breaks apart the crystalline solid.
     e.   both b and c occur.

2. Circle the letter of the most correct answer. The most important factor affecting the texture of an igneous rock is

    a. chemical composition of the magma or lava.
    b. pressure.
    c. rate of cooling.
    d. amount of space available for crystals to fill.

3. Circle the letter of the most correct answer. The most important factor affecting the mineral makeup of an igneous rock formed from a fluid magma is

    a. chemical composition of the magma or lava.
    b. pressure.
    c. rate of cooling.
    d. amount of space available for crystals to fill.

4. Circle the letter of the most correct answer. A coarse-grained igneous rock composed mainly of feldspar and amphibole mineral crystals in roughly equal proportions is called

    a. andesite.
    b. granite.
    c. gabbro.
    d. diorite.
    e. peridotite.

5. Circle the letter of the most correct answer. The proper sequence in which ferromagnesian silicate minerals crystallize from a cooling magma is

    a. amphibole, pyroxene, biotite, olivine.
    b. pyroxene, amphibole, olivine, biotite.
    c. biotite, amphibole, pyroxene, olivine.
    d. olivine, pyroxene, amphibole, biotite.

6. Circle the letter of the most correct answer. A volcanic rock consisting of abundant angular rock fragments is said to have what kind of texture?

    a. pyroclastic
    b. phaneritic
    c. porphyritic
    d. pegmatitic
    e. aphanitic

7.   For each of the following statements, write T for *true* or F for *false* on the space provided to indicate its validity.

(a) _____ Andesite is an extrusive igneous rock.
(b) _____ Cooling and crystallizing magma reverses the events of melting.
(c) _____ Tuff is a rock composed of volcanic ash.
(d) _____ In a cooling magma, Na-rich plagioclase forms before Ca-rich plagioclase.
(e) _____ Magmatic differentiation can be caused by crystal settling.
(f) _____ Pyroxene is a mineral of the continuous reaction series.
(g) _____ Granite is a mafic rock type.

CHAPTER 4: VOLCANOES AND OTHER IGNEOUS ACTIVITY

LEARNING OBJECTIVES

After reading and studying this chapter, you should be able to

1.   List and describe three factors that determine the nature of a volcanic eruption.

2.   Describe how temperature, composition, and gas content affect the viscosity of magma.

3.   List three categories of materials that may be emitted during a volcanic eruption.

4.   Contrast pahoehoe, aa, and pillow lava.

5.   Name and define the three main types of volcanoes as to size, shape, and eruptive style.

6.   Name a prominent example of each of the three main types of volcanoes.

7.   Distinguish between a crater and a caldera. Using Crater Lake as an example, describe the formation of a caldera.

8.   Discuss the possible effects of volcanism on climate.

9.   Describe the three criteria used to classify plutons.

10. List four different kinds of intrusive igneous bodies and describe each type in terms of the criteria used to classify plutons.

11. Describe how a batholith is emplaced.

12. Relate the distribution of volcanic activity to the plate tectonics model.

13. Define and understand the key terms listed at the end of the chapter.

VOCABULARY REVIEW

1.   The viscosity of a liquid is a measure of its _____.

2.   Basaltic lava flows with smooth surfaces, resembling braids of rope, are called (a) _____ flows. Basaltic lava flows with sharp, spiny surfaces and jagged blocks are called (b)_____ flows.

3.   _____ lava flows form in water.

4.   Dust, ash, and bombs ejected from volcanoes are collectively called (a)_____ material. If water mixes with such material, it may form a mudflow called a(n) (b)_____.

5.   A steep-walled depression located at the summit of a volcano is called a(n) (a)_____. If the width of such a depression is much greater than its depth, then it is called a(n) (b)_____.

6.   A simple volcano is a conical mountain with a steep-walled depression at its summit. The central depression occurs over a central vent; however, eruptions from secondary vents located on the flanks of a volcano may build (a)_____. Secondary vents that emit only gases are called (b)_____.

7.   Very broad, slightly domed volcanoes built from successive basaltic lava flows are called (a)_____, whereas smaller volcanoes made of pyroclastic material are called (b)_____.

8.   Bulbous masses of congealed magma in the vents of volcanoes act like giant plugs. These masses are called _____.

9.   Volcanoes composed of alternating layers of pyroclastic material and cooled lava flows are called (a)_____ or (b)_____.

10. A fiery cloud of hot gases and ash erupted violently from a volcano is called a(n) _____.

11. A cylindrical body of rock standing above the surrounding terrain after the rest of the volcano has been eliminated by erosion is called a _____.

12. The tube-like conduits of magma that "feed" volcanoes from depths of 200 kilometers or more inside of the earth are called _____.

13. Violent eruptions from linear volcanic vents are called (a)_____ eruptions. The extensive, very fluid basaltic lava flows that may result from such eruptions are known as (b)_____. Silica-rich lavas and associated volcanic ash may also flow violently and explosively from such eruptions, in which case masses of mixed ash and pumice move away from the vents at high speeds and high temperatures. Such hot, rapidly flowing masses of ash and pumice are called (c)_____.

14. Igneous bodies that cut across layers of sedimentary rock are said to be (a)_____, but those that form between layers of sedimentary rock are said to be (b)_____.

15. Remnants of country rock that have been incorporated into batholiths are called _____.

16. The gaseous components of magma (consisting mostly of water) are called _____.

17. Lava flows often contain tunnels that were horizontal conduits carrying lava from the volcanic vent to the lava flow's leading edge. These tunnels are called _____.

18. Volcanic eruptions from linear vents are called _____ eruptions.

19. Volcanism at a convergent plate boundary results in a linear or curved chain of volcanoes called a (a) _____. When volcanism occurs within a plate, it is called (b) _____.

APPLYING WHAT YOU HAVE LEARNED

1. What are hot spots, and how do they form? _____
_____
_____
_____
_____
_____

2. How does silica content affect a magma's viscosity? _____
_____
_____
_____

3. How does gas content affect a magma's viscosity? _____
_____
_____
_____

4. Study the three cross sections of volcanic cones presented below. On the blank provided below each profile, write shield, composite, or cinder to indicate which type of volcanic cone the profile represents.

(a) _____   (b) _____   (c) _____

5. Why is the circum-Pacific region called the Ring of Fire? _____
_____
_____
_____
_____

6.   Study the volcano illustrated below. What kind of volcano is it (in terms of how it formed)?
(a)_____ How did it form? (b) _____
_____
_____

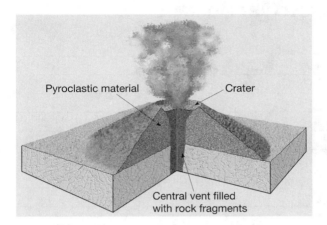

7.   On the blanks provided, write the name of the feature labeled. On the same blank, indicate
with capital letters whether the feature is concordant (C) or discordant (D) and whether the
feature is tabular (T) or massive (M).

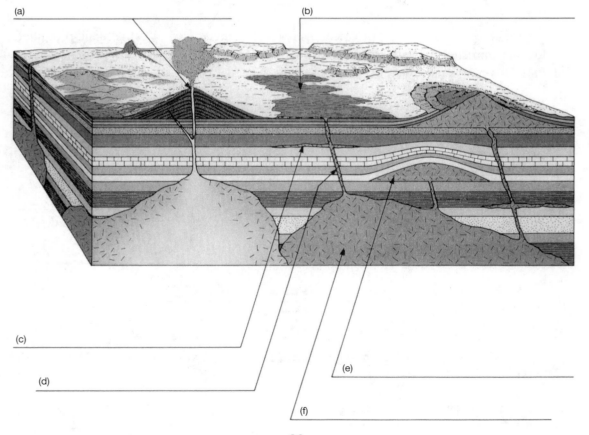

8.   What kind of lava flow is pictured below (based on the way the lava looks)? _____
_____

(Photo by J. D. Griggs, U.S. Geological Survey)

9.   Imagine that you are walking around the flanks of a volcano and that there are walnutsized fragments of basalt lying about that appear to have been ejected from the volcano. What would you call these fragments? _____

10. What are the three main factors that determine the nature of volcanic eruptions?
_____
_____
_____

11. What were the earliest indications that Mount St. Helens was going to erupt in 1980?
(a)_____
_____

What caused these indications? (b)_____
_____
_____
_____
_____

12. How can you distinguish a sill from a buried lava flow?_____
_____
_____
_____

ACTIVITIES AND PROBLEMS

1. Explain how are batholiths emplaced._____
    _____
    _____
    _____
    _____

2. What criteria are used to classify plutons? _____
    _____
    _____
    _____

3. Why are volcanoes associated with highly viscous magmas more explosive than volcanoes
    associated with fluid basaltic magmas? _____
    _____
    _____
    _____

REVIEW EXAM

1. Examine the map
    at the right and
    note the area of
    the map that is
    underlain by
    basalt (labeled).

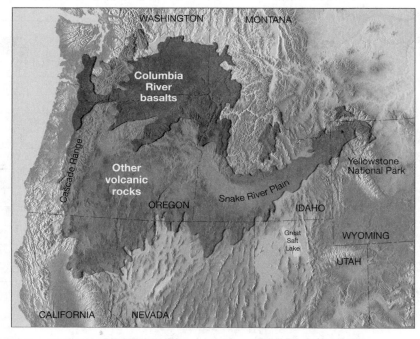

(After U.S. Geological Survey)

What is the name given to these particular basalts? (a)_____
How were they formed? (b)_____
_____

_____

_____

Suggest a reason why they formed. (c)_____

_____

_____

_____

2. Circle the letter of the most correct answer. The greatest volume of igneous rock is produced

   a. on and around the Hawaiian Islands.
   b. in subduction zones.
   c. along the oceanic ridge system.
   d. at hot spots.

3. The teardrop-shaped rock was one of many found lying about the flanks of a volcano. This one is about 8 cm long. What is the name given to such rocks? (a)_____ How do they form? (b)_____

   _____

4. For each of the following statements, write T for *true* or F for *false* on the space provided to indicate the validity of the statement.

   (a) _____ In near-surface environments, rocks of granitic composition begin to melt at about 800°C.

(b) _____ The average geothermal gradient in the upper crust of the Earth is about 1°C per kilometer.

(c) _____ Volcanoes of the Andes Mountains are mainly rhyolitic in composition.

(d) _____ The volcanoes that form the Hawaiian Islands are mainly shield volcanoes.

(e) _____ Mount St. Helens is located over a hot spot.

(f) _____ The volume of material extruded from Mount St. Helens in 1980 was exceptionally great in comparison to other volcanic eruptions of volcanoes throughout the world.

(g) _____ Numerous fissure eruptions occur along oceanic ridges.

(h) _____ Dikes are commonly formed along sedimentary bedding surfaces.

(i) _____ Granitic batholiths occur along the western margin of North America.

5. What is the name of the dark piece of rock that is located within the light-colored igneous rock pictured below? (a)_____ How did this rock (both light and dark portions) form? (b)_____

_____

_____

6. Fill in the blanks with the correct names of the labeled features of this volcano.

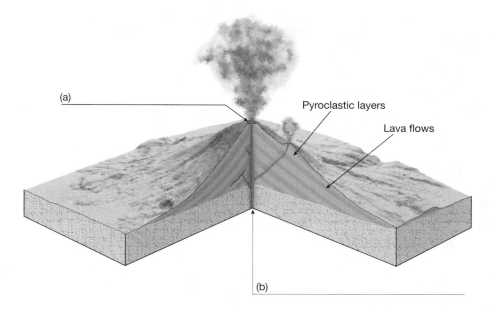

(a)

Pyroclastic layers

Lava flows

(b)

7. Study the illustrations below, showing three final events that formed Crater Lake, Oregon. Beside each illustration, write what happened (in your own words).

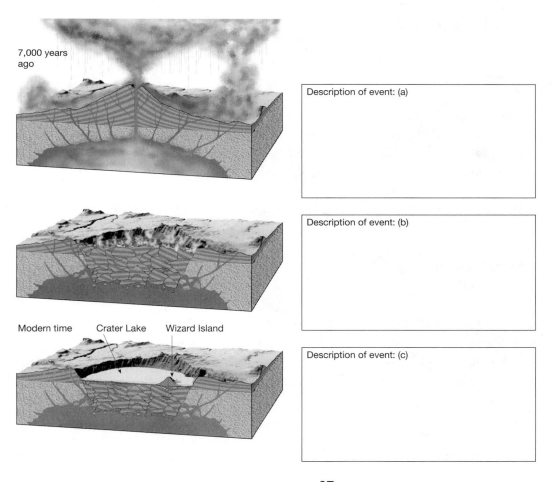

7,000 years ago

Modern time   Crater Lake   Wizard Island

Description of event: (a)

Description of event: (b)

Description of event: (c)

37

8.	Circle the letter of the most correct answer. As a lava flow gradually cools, contracts, and fractures, it can form a fracture pattern called

	a.	xenoliths.
	b.	nuée ardente.
	c.	parasitic cones.
	d.	columnar joints.

9.	Circle the letter of the most correct answer. The Yellowstone hot spot is an example of

	a.	intraplate volcanism.
	b.	volcanism at a convergent plate boundary.
	c.	flood basalts.
	d.	nuée ardente.

10.	Circle the letter of the most correct answer. The eruption of Vesuvius destroyed Pompeii in 79 a.d. by

	a.	burying Pompeii with lava.
	b.	burying Pompeii with ash and other pyroclastics.
	c.	melting Pompeii in a fissure eruption.
	d.	burying Pompeii in a giant lahar.

11.	Circle the letter of the most correct answer. The global atmospheric effect of a volcanic eruption depends mostly on

	a.	the volume of ash emitted during the volcanic eruption.
	b.	the volume of sulfuric acid aerosols produced as a result of the volcanic eruption.
	c.	how explosive the volcanic eruption was.
	d.	the volume of lava emitted during the volcanic eruption.

12.	Circle the letter of the most correct answer. Underwater volcanic eruptions of lava cause

	a.	aa lava flows.
	b.	pahoehoe lava flows.
	c.	pillow lava flows.

13.	Circle the letter of the most correct answer. Dikes are
	a.	discordant.
	b.	concordant.
	c.	massive.
	d.	both b and c.

14. Circle the letter of the most correct answer. The largest igneous intrusive bodies are

    a.   laccoliths.
    b.   dikes.
    c.   batholiths.
    d.   xenoliths.

CHAPTER 5: WEATHERING AND SOIL

LEARNING OBJECTIVES

After reading and studying this chapter, you should be able to

1. Describe the effects of external and internal processes on Earth's landscapes.

2. List and describe the three main external processes that transform rock into sediment.

3. List and describe four main types of mechanical weathering.

4. List and describe the three main types of chemical weathering.

5. Discuss the origin of carbonic acid and its role in rock weathering.

6. List and discuss three factors that influence the type and rate of rock weathering.

7. Sketch, label, and briefly describe a soil profile.

8. List and describe five basic controls of soil formation.

9. Explain what is meant by "acid precipitation" and why people are concerned about it.

10. Describe the difference between pedalfers, pedocals, and laterites.

11. Name and describe three factors that influence rates of soil erosion.

12. Describe the conditions that led to the Dust Bowl of the 1930s.

13. Explain why clearing of tropical and subtropical rain forests does not yield productive farmland.

14. Define and understand the key terms listed at the end of the chapter.

VOCABULARY REVIEW

1. The disintegration and decomposition of rock at or near the Earth's surface is called
    _____.

2. When water works its way into cracks and void spaces in rocks it may freeze. Upon freezing it expands and breaks the rock apart. This process is called _____.

3. The incorporation and transportation of Earth material by water, wind, or ice is called
    _____.

4. (a)_____is the process of decomposing rock, whereas
    (b)_____is the process of fragmenting and disintegrating rock.

5. The transfer of rock material downslope under the influence of gravity is called
    _____.

6. The reaction of any substance with water is called _____.

7. The process by which rocks weather to a more rounded shape is called
_____.

8. A layer of rock and mineral fragments produced by weathering is called
_____.

9. Rock fractures that form a definite pattern are called _____.

10. _____ is the process by which rusting occurs.

11. The process by which materials dissolve is called _____.

12. A layer of rock and mineral fragments, organic matter, water, and air covering the land is called _____.

13. Piles of fragmented rock at the base of steep rock outcrops are called _____.

14. Large igneous bodies beneath the surface of the Earth are under great pressure from overlying rocks. When the overlying rocks are stripped away by weathering, erosion, and mass wasting, then layers of rock may break away from the exposed surface of the igneous body. Such a process is known as (a)_____ and the large exposed surfaces of such structures are called (b)_____.

15. Soils are composed of layers called (a)_____. A vertical cross section through all of these layers is called a(n) (b)_____

16. The process of washing fine-grained soil components from a soil, as water percolates downward through it, is called _____.

17. A layer of decayed remains of plants and animals (organic matter) is called
_____.

18. Volcanism and mountain building are examples of (a)_____ processes.
Weathering, mass wasting, and erosion are examples of (b)_____ processes.

19. Depletion of soluble materials from the upper layers of soil is called
_____.

20. The source of weathered mineral matter from which soils develop is called
_____.

21. The "true soil" is called the _____.

22. Soils rich in iron oxides and clay minerals are called (a)_____. Soils rich in calcium carbonate are called (b)_____.

23. The kind of weathering that forms unusual rock formations and landforms due to variations in mineral makeup, degree of jointing, and exposure to the elements, is called _____ _____ weathering.

APPLYING WHAT YOU HAVE LEARNED

1. What is spheroidal weathering, and how does it occur?_____
   _____
   _____
   _____

2. What is acid precipitation, and why are people concerned about it? _____
   _____
   _____
   _____
   _____

3. The photo below was taken on a hilltop in Joshua Tree National Monument, California, by E. J. Tarbuck. What are these types of cracks called? (a)_____.
   What processes are enlarging these cracks? (b)_____
   _____
   What process is causing the blocks of rock to become rounded? (c)_____
   _____

4.   What is caliche, and how does it form? _____
      _____
      _____
      _____

5.   Which soil horizon is composed mostly of weathered bedrock? _____
      _____

6.   What is oxidation, and how does it occur? _____
      _____
      _____
      _____

7.   What is the feature labeled "A" in the photograph below? (a)_____
      _____ How did it form? (b)_____
      _____
      _____

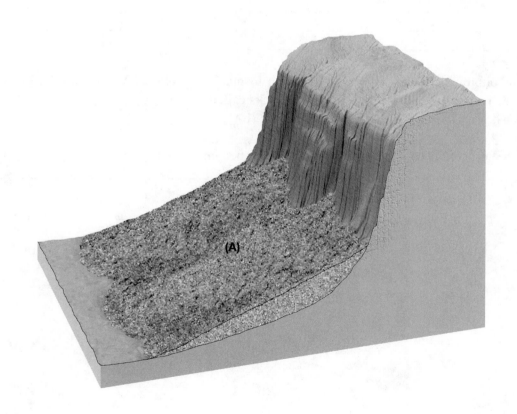

ACTIVITIES AND PROBLEMS

1. Which item would chemically weather fast when exposed to an acidic solution: calcite sand weighing 10 grams or a single chunk of calcite weighing 10 grams?
   (a)_____Why? (b)_____
   _____

2. Suggest a rule relative to Bowen's reaction series that allows you to predict the order in which silicate minerals decompose at the earth's surface. _____
   _____
   _____

3. Carefully examine the illustration above and the features labeled A through D. Give a letter (A, B, C, or D) for each answer below.

   At which location would the thickest residual soil develop?
   (a)_____
   At which location would the greatest amount of frost wedging occur?
   (b)_____
   At which location would the thickest transported soil develop? (c)_____
   At which location would there be the most unconsolidated deposits? (d)_____

4. How does hydrolysis cause the decomposition of minerals? _____
   _____
   _____

5.   How does climate affect rock weathering? _____
_____
_____
_____

REVIEW EXAM

1.   Name and describe four types of mechanical weathering. _____
_____
_____
_____

2.   In general, how does time affect soil development? _____
_____
_____
_____

3.   How and why does the topography of arid regions generally differ from the topography of humid regions? _____
_____
_____

4.   How does carbonic acid form? _____
_____
_____

5.   Complete the chart below.

SOIL TYPE	CLIMATE	SUBSOIL IS ENRICHED IN	TOPSOIL COLOR
pedalfers	Temperate Humid	(a)	dark to light in color
pedocals	Temperate Dry	(b)	whitish colors
laterites	(c)	No subsoil present: Removed by leaching	(d)

6.   List five basic controls of soil formation. _____

_____

_____

_____

_____

7.   Circle the letter of the most correct answer. In which of the following areas are laterites best developed?

   a.   the north pole
   b.   southwestern U.S.
   c.   northeastern U.S.
   d.   Brazil

8.   Circle the letter of the most correct answer. In which of the following areas would a pedocal be most likely to form?

   a.   the north pole
   b.   southwestern U.S.
   c.   northeastern U.S.
   d.   Brazil

9.   Circle the letter of the most correct answer. Which item below is *not* a chemical weathering process?

   a.   oxidation
   b.   hydrolysis
   c.   thermal expansion
   d.   dissolution

10.   Circle the letter of the most correct answer. Which item below is *not* a mechanical weathering process?

   a.   frost wedging
   b.   biological activity
   c.   hydrolysis
   d.   unloading

11.   Circle the letter of the most correct answer. Which item below does *not* influence rates of soil erosion?

   a.   climate
   b.   type of bedrock
   c.   slope
   d.   type of vegetation

12. Explain why productive farmland is not produced by the clearing of tropical and subtropical rain forests. _____

_____

_____

_____

13. What conditions led to the Dust Bowl of the 1930s? _____

_____

_____

_____

14. For each of the following statements, write T for *true* or F for *false* on the space provided to indicate its validity.

(a) _____ Calcite dissolves only in strongly acid solutions.
(b) _____ A soil developed from beach sands would be an example of a transported soil.
(c) _____ The B horizon of a soil profile is known as the "subsoil."
(d) _____ The organic-rich layer at the very top of a soil profile is called the organicum.
(e) _____ Steep slopes typically have soils with well-developed soil profiles.

15. Fill in the blanks with the correct names of the soil "layers" that are labeled.

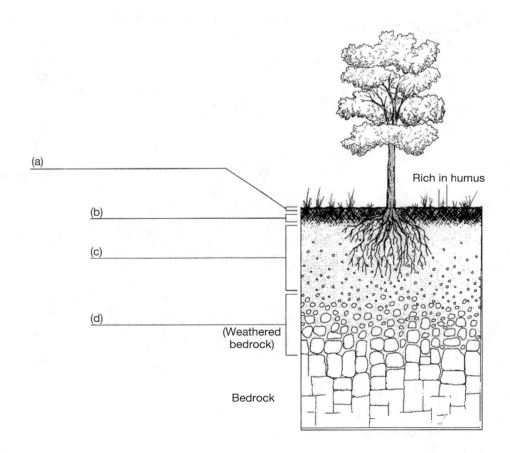

(a) _____

(b) _____

(c) _____

(d) _____

Rich in humus

(Weathered bedrock)

Bedrock

48

CHAPTER 6: SEDIMENTARY ROCKS

LEARNING OBJECTIVES

After reading and studying this chapter, you should be able to

1. Explain how weathering, erosion, deposition, diagenesis, and lithification contribute to the formation of sedimentary rocks.

2. Name and describe the two ways that most sedimentary rocks become lithified.

3. Briefly explain why sedimentary rocks are important in the study of Earth history.

4. Contrast the two main groups of sedimentary rocks.

5. List and briefly describe the origin and chief characteristics of the four detrital sedimentary rocks. List and briefly describe the origin and chief characteristics of the five common chemical sedimentary rocks.

6. Explain how sedimentary rocks are related to the carbon cycle.

7. List six different sedimentary structures and indicate which one is the most common.

8. List two conditions that favor the preservation of organisms as fossils.

9. List the minerals that precipitate from a body of sea water as it evaporates and the order in which they precipitate.

10. Briefly describe how facies are related to sedimentary environments.

11. Explain how seafloor sediments can reveal how the Earth's climate has changed.

12. Define and understand the key terms listed at the end of the chapter.

VOCABULARY REVIEW

1. (a)_____is any mass of particles of Earth materials that have been weathered, eroded, transported, and deposited. (b)_____ is the process by which such a deposit of loose particles is transformed (hardened) into sedimentary rock.

2. Some sedimentary rocks are composed of transported solid particles that were derived from both mechanical and chemical weathering of earth materials. Such rocks are called _____ sedimentary rocks.

3. Other sedimentary rocks are composed of interlocking crystals that were precipitated from solutions containing dissolved substances or are composed of materials formed during the life processes of animals and plants. Such rocks are called _____ sedimentary rocks.

4. Sediments accumulate in layers called (a)_____. These layers are separated from one another by surfaces called (b)_____ _____.

5. As sediments accumulate through time, the weight of overlying material compresses the deeper sediments. This process is called _____.

6. Some sedimentary layers change from relatively large particles at their bases to finer particles at their tops. These layers are called (a)_____. Other layers are deposited at a steep angle from the horizontal. Such layering is called (b)_____.

7. _____ are small waves of sand that develop on the surface of a sediment layer by the action of moving water or wind.

8. _____ form when wet mud dries out and shrinks because it is exposed to air.

9. The degree to which particle size is the same among grains in a sedimentary rock is referred to as _____.

10. (a)_____ is the ability of some rocks to split into thin layers along well-developed, closely spaced planes. The fine-grained sedimentary rock in which this property is best developed is (b)_____.

11. Remains or traces of prehistoric life are called _____.

12. When water containing dissolved substances evaporates from the floor of an enclosed basin, a light-colored crust of mineral crystals is left behind on the surface of the ground. Such mineral-encrusted land surfaces are called (a)_____ and the mineral deposits are called (b)_____.

13. The rock-forming process by which sediments are "glued" together by chemical precipitates is the process of (a)_____. The chemically precipitated material that acts as a "glue" in this process is called (b)_____.

14. Sedimentary rocks consisting of discrete fragments of broken particles that have been compacted and cemented together display a(n) _____ texture.

15. Sedimentary rocks composed of mineral crystals that form an interlocking pattern have a(n) _____ texture.

16. A(n) _____ is a rock unit having distinctive characteristics that reflect the conditions of a particular environment in which it formed.

17. A geographic setting where sediment is accumulating is called a(n)
_____.

18. _____ is the term that refers to all of the physical, chemical, and biological changes that occur after sediment is deposited and during and after it is lithified.

APPLYING WHAT YOU HAVE LEARNED

1. What are fossils? _____
   _____
   _____

2. Name and describe five different types of sedimentary structures. _____
   _____
   _____
   _____
   _____
   _____
   _____
   _____
   _____
   _____
   _____
   _____
   _____

3. What is a facies? _____
   _____
   _____
   _____

4. What is lithification? _____
   _____
   _____

5. What is the compositional difference between limestone and dolomite (dolostone)? _____
   _____
   _____
   _____

6. What is chert? _____
   _____
   _____

7. What kind of sedimentary structure is pictured in the photograph below? (a)_____
_____ Explain how this kind of sedimentary structure forms. (b)_____
_____
_____

(Photo by E. J. Tarbuck)

8. Carefully observe each of the photographs below of sedimentary rocks. To the right of each
sample, name the rock and describe its origin (how it formed).

SAMPLE   NAME AND ORIGIN OF ROCK SAMPLE

(a) _____

(b) _____

(c) _____

9.   Give the particle size in millimeters for each kind of sediment:

Gravel   (a) _____ mm
Sand   (b) _____ mm
Mud   (c) _____ mm

10.   What are detrital sedimentary rocks? _____
_____
_____
_____

11.   What are chemical sedimentary rocks? _____
_____
_____
_____

12.   Name two ways that carbon is stored in sedimentary rocks, and describe how it got stored
there. _____
_____
_____
_____

13. How does carbon move from sedimentary rocks (lithosphere) to the hydrosphere?

_____

_____

_____

_____

ACTIVITIES AND PROBLEMS

1. Complete the chart below.

Name of sedimentary rock	What texture is generally present in this kind of rock?	What is the size of particles in this kind of rock	Is the rock detrital, chemical, or biochemical
Mudstone	Clastic	less than 1/16 mm	detrital
Rock salt	(a)	(b)	(c)
Arkose	(d)	(e)	(f)
Conglomerate	(g)	(h)	(i)
Limestone	(j)	(k)	(l)
Rock gypsum	(m)	(n)	(o)

2. What is sorting, and how does it occur? _____

_____

_____

_____

_____

3. How does coal form? _____

_____

_____

_____

_____

_____

_____

4.   How does cementation occur? _____
_____
_____

REVIEW EXAM

1.   Name and describe the two main ways that sedimentary rocks become lithified. _____
_____
_____
_____
_____

2.   Briefly explain how sedimentary rocks form relative to the rock cycle. _____
_____
_____
_____
_____
_____

3.   Why are fossils important? _____
_____
_____
_____
_____

4.   What are reefs, and how are they formed? _____
_____
_____
_____
_____

5.   Briefly note why sedimentary rocks are important in the study of Earth history. _____
_____
_____
_____
_____
_____

6.   Explain how seafloor sediments can reveal how Earth's climate has changed. _____
_____
_____
_____

7. The pictures below are cross sections through layers of sediment. Name the sedimentary structure that is present in each cross section.

(a)_____   (b)_____

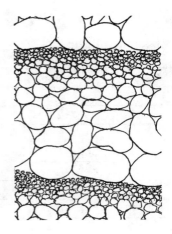   

8. Complete the chart below.

Percent of sea water evaporated	Minerals precipitated from the remaining water
0	None
50-75	(a)
90	(b)
98	potassium and magnesium salts

9. Circle the letter of the correct answer. The two conditions that most favor the preservation of an organism as a fossil are

    a.   presence of oxygen and slow burial.
    b.   presence of many soft parts and rapid burial.
    c.   presence of hard parts and slow burial.
    d.   presence of hard parts and rapid burial.

10. For each of the following statements, write T for *true* or F for *false* on the space provided to indicate its validity.

(a) _____ Ripple marks could form in a desert.

(b) _____ Limestone can have a clastic or nonclastic texture.

(c) _____ Chalk is a kind of limestone.

(d) _____ If a detrital sedimentary rock is composed of angular, pebble-sized particles, then the rock is conglomerate.

(e) _____ As ocean water evaporates, gypsum (hydrous calcium sulfate) forms before halite (sodium chloride).

CHAPTER 7: METAMORPHIC ROCKS

LEARNING OBJECTIVES

After reading and studying this chapter, you should be able to

1.   List three agents of metamorphism and briefly describe the effects of each agent.

2.   Contrast rock cleavage and schistosity.

3.   Explain why some metamorphic rocks are nonfoliated.

4.   Describe three settings in which metamorphism most often occurs.

5.   Compare and contrast marble and quartzite.

6.   Briefly describe the textural and mineralogical differences among slate, mica schist, and gneiss.

7.   Describe the use of index minerals in the interpretation of metamorphic grades.

8.   Explain why migmatites are both igneous and metamorphic.

9.   Describe how black smokers and metallic ore deposits form along ocean ridge systems.

10.   Discuss the relationship between metamorphism and plate tectonics.

11.   Explain briefly how impact metamorphism and tektites may be related.

12.   Define and understand the key terms listed at the end of the chapter.

VOCABULARY REVIEW

1.   Metamorphism takes place on two different scales. When extensive portions of the crust are metamorphosed, then the rocks are said to have undergone
(a)_____. When rock is metamorphosed locally, because of direct contact with a mass of magma, it is said to have undergone
(b)_____.

2.   Metamorphic rocks are produced or transformed from preexisting igneous, metamorphic, or sedimentary rocks called _____.

3.   During compressional processes associated with mountain building or the burial of rocks, the rocks are subjected to force, or _____.

4.   Extensive, relatively flat areas of continents that are underlain by old metamorphic and igneous rocks are called _____.

5.   Hot ion-rich water released from bodies of magma is called
      a_____. It causes chemical alteration of rocks, which is
      known as (b) _____ metamorphism.

6.   During contact metamorphism, a zone of alteration (or halo) forms in the outermost edge of
      the emplaced body of magma. Such a zone, or halo, is called
      a(n)_____.

7.   _____ metamorphism is a low grade of metamorphism that
      occurs beneath thick accumulations of sedimentary strata in a subsiding basin.

8.   Directional stresses can cause a rock body to break into slabs, which slide past one another.
      This is called _____.

9.   Certain minerals are good indicators of the specific kind of metamorphic environment in
      which they formed. These minerals are called _____.

10.   (a)_____results whenever mineral crystals and structural fea-
      tures of a rock body are brought into parallel alignment, giving the rock body a layered
      texture. Rock bodies that have such a parallel alignment of mineral crystals and structural
      features can be easily split along their layered texture. Such splitting is known as
      (b)_____ or (c)_____.

11.   When mica mineral crystals in slate are recrystallized to larger sizes, the rock develops a
      platy, or scaly, type of foliation called (a)_____. This type of
      foliation is best developed in the metamorphic rock called (b)_____.

12.   Metamorphic rocks that do not have foliated texture are called
      _____ metamorphic rocks.

13.   _____ texture is present in metamorphic rocks when large
      mineral crystals are set in a fine-grained mass of other minerals.

14.   _____ is the name of a metamorphic rock with banding caused
      by differences in mineralogy.

15.   _____ is a nonfoliated, crystalline metamorphic rock composed
      of calcite.

16.   _____ is a very hard, nonfoliated metamorphic rock formed
      from quartz sandstone.

17.   _____ is a very fine-grained, foliated metamorphic rock with
      excellent rock cleavage.

18. _____ are rock bodies of dark, foliated, highly metamorphosed rock having bands of light-colored crystalline igneous rock.

19. _____ is the size, shape, and distribution of particles that constitute a rock.

APPLYING WHAT YOU HAVE LEARNED

1. What is foliation? _____
   _____
   _____

2. What is hornfels, and how does it form? _____
   _____
   _____
   _____

3. Describe how black smokers and metallic ore deposits form along ocean ridge systems.
   _____
   _____
   _____
   _____
   _____
   _____

4. How are slaty cleavage and schistosity related? (a)_____
   _____
   _____
   _____

   How are they different? (b)_____
   _____
   _____
   _____
   _____

5. Write a brief description for each of the metamorphic rock types listed below.

   (a)   quartzite: _____
   (b)   slate: _____
   (c)   gneiss: _____
   (d)   phyllite: _____
   (e)   marble: _____

ACTIVITIES AND PROBLEMS

1. How do migmatites form? _____
   _____
   _____
   _____

2. In terms of plate tectonics, where would you expect to find zones of high-pressure metamor-phism? _____
   _____
   _____
   _____

3. Carefully examine the photograph below of a rock sample. The sample is shown at its actual size. Also think of all of the different parts of the rock cycle that have to do with the formation of this rock.

(Photo by E. J. Tarbuck)

Now list all the steps of the rock cycle that were involved in the formation of this rock as it appears here and now. Start with step a, that is, weathering of preexisting rocks at the surface of the Earth.

(a)   Weathering of preexisting rocks occurred at the surface of the Earth. _____
(b)   _____
(c)   _____
(d)   _____
(e)   _____
(f)   _____

4. Explain briefly how impact metamorphism and tektites may be related. _____

_____
_____
_____
_____
_____
_____

5. Describe three settings in which metamorphism most often occurs. _____

_____
_____
_____
_____
_____

6. Carefully examine the photograph below of a metamorphic rock sample. The sample is shown at its actual size.

(Photo by E. J. Tarbuck)

Is this rock foliated or nonfoliated? (a) _____

What is the name of this metamorphic rock? (b)_____

Consider how this rock got its texture. Was it compressed from top to bottom or from side to side? (c)_____

7. Briefly describe the process that has led to the formation of intricate folding in solid rocks that now crop out at the surface of the Earth. _____

_____

_____

_____

8. Name three agents of metamorphism. _____

_____

_____

_____

9. What is confining pressure? _____

_____

_____

10. The rock pictured below is composed of large dark-colored garnet crystals set in a mass of small muscovite mica crystals that are aligned parallel to one another. Is this rock foliated or nonfoliated? (a)_____ What is the name of this kind of metamorphic rock? (b)_____ What other two names can be used to describe the textures of this rock? (c) _____

(Photo by E. J. Tarbuck)

11.   The rock pictured below is composed of calcite. Is this rock foliated or nonfoliated?
        (a)_____ What is the name of this kind of metamorphic rock?
        (b)_____

REVIEW EXAM

1.   When does the process of metamorphism begin, and when does it end? _____
        _____
        _____
        _____
        _____

2.   What effect(s) do hydrothermal solutions have on a body of rock?_____
        _____
        _____
        _____
        _____

3. Carefully examine the cross section below of a subduction zone, and note the area labeled A, B, and C.

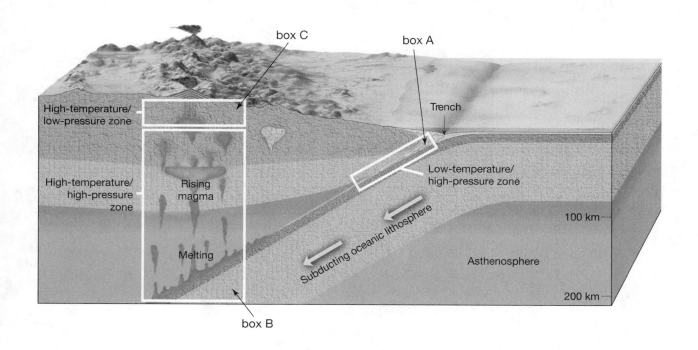

At which location (A, B, or C) would a metamorphic regime occur that is characterized by high temperature and high pressure? (a)_____

At which location (A, B, or C) would a metamorphic regime occur that is characterized by high temperature and low pressure? (b) _____

At which location (A, B, or C) would a metamorphic regime occur that is characterized by low temperature and high pressure? (c) _____

At which location (A, B, or C) would blueschist form? (d) _____

4. Describe the metamorphic conditions that produce hornfels. _____
_____
_____
_____

5. On the blank provided beside each index mineral or rock type listed below, write *low, inter-mediate*, or *high* to indicate the grade of metamorphism that it represents.

    (a)   migmatite: _____

    (b)   chlorite: _____

    (c)   sillimanite: _____

    (d)   slate: _____

    (e)   schist: _____

    (f)   gneiss: _____

    (g)   garnet: _____

6. How do serpentine-rich and talc-rich rocks form on the seafloor? _____

    _____

    _____

    _____

    _____

7. Circle the letter of the correct answer. In which of the following ways is the greatest volume of metamorphic rocks produced?

    a.   contact metamorphism

    b.   regional metamorphism

    c.   metamorphism by hydrothermal solutions

    d.   movements along faults

8. For each of the following statements, write T for *true* or F for *false* on the space provided to indicate its validity.

    (a) _____ In general, recrystallization causes new kinds of minerals to form.

    (b) _____ Low-grade metamorphism makes rocks more compact, so they are also made more dense.

    (c) _____ Phyllite represents a grade of metamorphism between that of schist and slate.

    (d) _____ The parent rock for marble is quartz sandstone.

    (e) _____ Gneiss has no foliations.

    (f) _____ Contact metamorphic rocks usually have a coarse-grained texture.

CHAPTER 8: GEOLOGIC TIME

LEARNING OBJECTIVES

After reading and studying this chapter, you should be able to

1. Describe two early methods that were used for dating the Earth and list some weaknesses of each.

2. Distinguish between relative and numerical dating.

3. Apply relationships of superposition, inclusions, and cross-cutting to solving problems of simple relative age dating.

4. Contrast disconformities, nonconformities, and angular unconformities.

5. Explain the concept of correlation.

6. State the principle of faunal succession and briefly discuss the importance of fossils in relative age dating.

7. List and briefly describe three common types of radioactive decay.

8. Calculate a numerical age given the half-life of a radioactive isotope and the ratio between the radioactive parent and the stable daughter product.

9. Briefly describe the use of carbon-14 and potassium-40 in radiometric dating. List some advantages and limitations of these isotopes in radiometric dating.

10. Discuss factors that could lead to unreliable radiometric dates.

11. Describe how the geologic time scale was constructed.

12. Briefly discuss the difficulties encountered in subdividing the Precambrian.

13. Understand what radon is, how it enters a home, and why it is a health hazard.

14. Explain the difficulties associated with assigning absolute dates to sedimentary strata.

15. Define the KT extinction event and evaluate what may have caused it.

16. Describe the value of tree rings as a record of time and environmental change.

17. Define and understand the key terms listed at the end of the chapter.

VOCABULARY REVIEW

1. _____ means that rocks or events are placed in their correct sequence, or order.

2. _____ pinpoint the time in history when some event took place.

3. The _____ states that in an un-deformed sequence of strata, each bed is older than the one above it and younger than the one below it.

4. The _____ states that when rocks are cut by another feature (fracture, fault, igneous intrusive body), the rocks are older than the cutting feature.

5. The _____ states that layers of sediment are generally deposited in nearly flat-lying position.

6. _____ are pieces of rock that are contained within another rock.

7. Strata are said to be (a)_____ when they have been deposited one on top of the other without interruption. Wherever there has been an interruption in the deposition of layers (or wherever erosion has removed previously formed rocks), there are breaks in the rock record that are called (b)_____.

8. The task of matching up rocks of similar ages among different regions is referred to as (a)_____. The (b)_____ states that fossil organisms (including plants) succeed one another in a definite and determinable order. Fossils that are associated with a particular span of geologic time are called (c)_____.

9. _____ occur where folded or tilted sedimentary rocks are separated from overlying strata that are flat-lying.

10. _____ are breaks that separate older metamorphic or intrusive igneous rocks from younger sedimentary strata.

11. _____ is another name for C-14.

12. The process by which nuclei of atoms spontaneously break apart, or decay, is called (a)_____. It is the basis for a type of absolute dating called (b)_____.

13. A(n) _____is an unconformity between parallel strata.

14. The largest standard division of geologic time is a(n) (a)_____, which can be subdivided into smaller standard subdivisions of time called (b)_____. On the other hand, the smallest standard divisions of geologic time are called (c)_____, which are commonly grouped together to form time divisions called (d)_____.

15. A radioactive isotope is referred to as a parent isotope, and it decays to form _____.

16. The time required for one-half of the nuclei of an unstable isotope to decay is called _____.

17. The era of ancient life is called the (a)_____ Era, the era of recent life is called the (b)_____ Era, and the era of middle life is called the (c)_____ Era.

18. The eon of visible life is called the _____ Eon.

19. The more than 4 billion years of time prior to the Cambrian Period is referred to as the (a)_____. It is commonly subdivided into eons from the oldest (b)_____ Eon to the (c) _____ Eon to the youngest (d)_____ Eon.

20. Geologists have divided geologic history into a chart of dated and named eons, eras, periods, and epochs. This chart is called the _____.

APPLYING WHAT YOU HAVE LEARNED

1. What is an unconformity? _____
_____
_____

2. State the law of superposition. _____
_____
_____
_____

3. State the principle of original horizontality. _____
_____
_____
_____

4. State the principle of cross-cutting relationships. _____
_____
_____
_____
_____

5. Why is it that any eon, era, period, or epoch of geologic history can be recognized on the basis of its fossils? _____
_____
_____

6. Imagine that the advertisement below appeared in your local newspaper. What are two scientific reasons why you should be suspicious about the validity of the ad? _____
_____
_____
_____

FOR SALE
World's oldest fossil!
•
100 million years old.
Age verified by
carbon dating.
•
Make Offer!

7. What is the task of correlation? _____
_____
_____

8. What is half-life? _____
_____
_____
_____

9. On the blank provided beside each geologic cross section below, write the name of the specific type of unconformity that is labeled with an arrow. The v-pattern indicates igneous rock. All other patterns are different types of sedimentary rocks

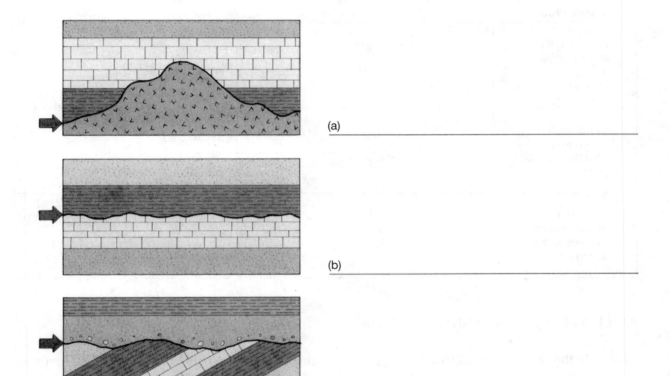

(a) _____

(b) _____

(c) _____

10. Fill in the chart below, which concerns common types of radioactive decay.

Atomic Process	Name of This Type of Radioactive Decay	Change in Atomic Number	Change in Atomic Mass
Emission of an alpha particle from the nucleus of an atom	(a)	(b)	(c)
Emission of beta particle from the nucleus of an atom	(d)	(e)	(f)
Capture of an electron (by the nucleus of the atom)	(g)	(h)	(i)

ACTIVITIES AND PROBLEMS

1. In the chart of radioactive decay presented below, how many half-lives have elapsed by time (a)?_____. How many half lives have elapsed by time (b)?_____.

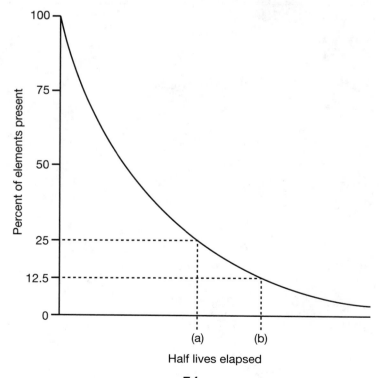

2. The radioactive parent isotope potassium-40 decays to the stable daughter product argon-40 and has a half-life of 1.3 billion years. If a sample of igneous rock is found to contain a ratio of 1/8 potassium-40 to 7/8 argon-40, then how old is the rock sample? _____

3. A chemist determines that the ratio of radioactive parent isotope to stable daughter isotope in a sample is 1:3. How many half-lives have transpired? _____

4. If any parent material has undergone 8 half-lives of decay, then how much of the parent material remains as a ratio of the original amount? (a)_____
What percentage of the original amount of parent material is this? (b)_____
_____ If any parent material has undergone 7 half-lives of decay, then how much of the parent material remains as a ratio of the original amount? (c)_____
_____ What percentage of the original amount of parent material is this?
(d)_____

   If measurement of the amount of a parent material involves an error of plus or minus 0.5%, would it be possible to distinguish whether that parent material has undergone 7 half-lives of decay versus 8 half-lives of decay? Explain. (e)_____
_____
_____
_____

5. Recall that the half-life of carbon-14 (radiocarbon) is 5,730 years. Why is it that radiocarbon dating can be used only to date carbon-rich materials less than 75,000 years old? _____
_____
_____
_____
_____

6. If a radioactive atom of uranium-234 emits two alpha particles, then what will be the atomic mass of the daughter product that is produced? (a)_____ If the same atomic number of uranium-234 is 92, then what will be the atomic number of the same daughter product? (b)_____

7. What is radon, why is it a health hazard, and how can it enter a home? _____
_____
_____
_____
_____
_____
_____
_____
_____
_____

8. Examine the two sections of strata below from locations X and Y. The patterns in the two sections represent types of sedimentary rocks. Give the letter of the rock unit at location Y that correlates with rock unit c of location X. (a)_____ Give the letter of the rock unit at location X that correlates with rock unit e of location Y. (b)_____

Location X

Location Y

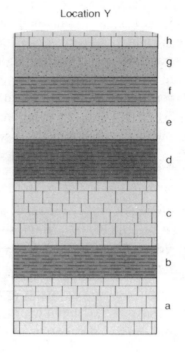

9. Examine the geologic cross section below. Which feature is the youngest? (a)_____ Which feature is the oldest? (b)_____

REVIEW EXAM

1. Give a step-by-step description of how an angular unconformity can develop, starting with step a and proceeding through additional steps as needed.

    (a)_____

    (b)_____

    (c)_____

2. Name two early methods that were used to estimate the numerical age of the Earth, and give reasons why both methods were not accurate. _____

    _____

    _____

    _____

    _____

3. The numerical age of the Earth is about (a)_____ billion years, which is the same as (b)_____ million years.

4. A(n) _____ is the specific name given to an unconformity formed between nonparallel strata.

5. How are inclusions used for relative age dating? _____

    _____

    _____

    _____

6. Why is the Precambrian not subdivided into detailed time divisions as is the Phanerozoic Eon? _____

    _____

    _____

    _____

7. In general, how was the geologic time scale constructed? _____

    _____

    _____

    _____

    _____

8. Give two distinctly different reasons why radiometric dates are sometimes unreliable (assuming that analytical laboratory techniques are totally reliable). _____

_____

_____

_____

_____

9. Circle the letter of the most correct answer. Numerical dating is

   a.   the process of placing events in correct sequence through time.
   b.   used to pinpoint the time in years when an event occurred.
   c.   determined with the law of superposition.
   d.   determined with radiometric dating techniques.
   e.   both a and c.
   f.   both a and b.
   g.   both b and d.

10. Circle the letter of the most correct answer. A beta particle is composed of

   a.   1 neutron.
   b.   1 electron.
   c.   1 electron and 1 proton.
   d.   1 proton and 1 neutron.
   e.   2 protons and 2 neutrons.
   f.   2 protons and 2 electrons.

11. Circle the letter of the most correct answer. Tree rings record information about

   a.   time.
   b.   environmental change.
   c.   alpha emissions.
   d.   beta emissions.
   e.   both a and b.
   f.   both c and d.

12. Circle the letter of the most correct answer. The eon of visible life is called the

   a.   Archean.
   b.   Phanerozoic.
   c.   Hadean.
   d.   Proterozoic.

13.   Circle the letter of the most correct answer. We are living in the

      a.   Cenozoic Era.
      b.   Paleozoic Era.
      c.   Mesozoic Era.

14.   Examine the illustration below (a cross section through the Earth as you might observe on the wall of a quarry or a road cut). Give two reasons why the granite must be older than the sedimentary layers. _____

_____

_____

_____

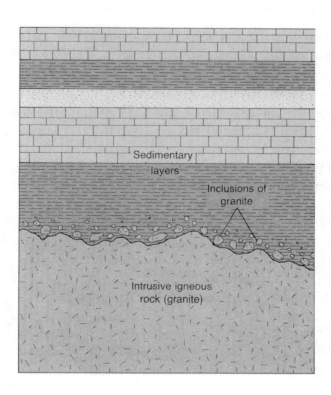

15. Fill in the chart below with the correct name of a geologic era of time.

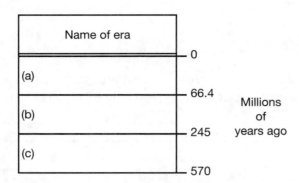

16. For each of the following statements, write T for *true* or F for *false* on the space provided to indicate the validity of the statement.

(a) _____ All isotopes of uranium are radioactive.

(b) _____ The Pennsylvanian Period is named for the state of Pennsylvania, where these rocks have produced much coal.

(c) _____ The Mesozoic Era is composed of the Tertiary, Jurassic, and Cretaceous periods.

(d) _____ John Wesley Powell led a pioneering expedition down the Colorado River.

(e) _____ Archbishop Ussher estimated that the Earth was about 3 million years old.

(f) _____ Lord Kelvin estimated the age of the Earth by assuming that it was once molten and that it had cooled to its present condition.

(g) _____ Steno made correlations based upon the principle of fossil succession.

(h) _____ William Smith is credited with recognizing the law of superposition and the principle of original horizontality.

(i) _____ The KT extinction marks the end of the Paleozoic Era.

CHAPTER 9 : MASS WASTING—THE WORK OF GRAVITY

LEARNING OBJECTIVES

After reading and studying this chapter, you should be able to

1. Describe the relationship of mass wasting processes to weathering, erosional processes, and landforms.

2. Explain the importance of gravity, water, angle of repose, vegetation, and earthquakes, as controls and triggers of mass wasting.

3. Describe the three factors commonly used to classify mass wasting processes.

4. Distinguish among fall, slide, and flow, relative to mass wasting.

5. Explain the difference between a slump and an earthflow.

6. Describe the special circumstances that produce a lahar.

7. Explain the difference between an earthflow and a mudflow.

8. Explain the mechanism that is responsible for creep and describe some visible effects of this process.

9. Describe the nature and distribution of permafrost and the mass wasting process that is unique to this zone.

10. Know that landslides are not confined to the land; submarine landslides also occur.

11. Define and understand the key terms listed at the end of the chapter.

VOCABULARY REVIEW

1. The downslope movement of rock, regolith, and soil under the influence of gravity is called _____.

2. The steepest angle at which the surface of a mass of loose particles remains stable is called the _____.

3. Any free-fall of detached pieces of rock, regolith, or soil is called a(n) (a) _____. Any movement of material along a well-defined surface is called a(n) (b) _____. Any movement of material acting as a viscous fluid is called a(n) (c) _____.

4. The most rapid of mass movements are termed _____.

5. The very slow, downhill movement of soil and regolith caused by freezing and thawing or wetting and drying is called _____.

6. Permanently frozen ground is called a(n) (a) _____.
   The form of creep that is common to regions having such permanently frozen ground is called a(n) (b) _____.

7. Slope failure involving downward sliding of masses of solid rock or unconsolidated material (regolith or soil) along curved, spoon-shaped surfaces is called

   _____.

8. Movements that occur when blocks of bedrock break loose and slide down a slope are called (a) _____. If such movements largely involve unconsolidated material, however, then they are termed (b) _____.

9. A(n) (a) _____ is a relatively rapid type of mass wasting that involves flowage of unconsolidated debris containing much water. If this type of mass wasting involves volcanic ash and debris on the slope of a volcano, then it is called a(n) (b) _____.

APPLYING WHAT YOU HAVE LEARNED

1. What is the difference between a slump and a rockslide? _____
   _____
   _____
   _____
   _____

2. Examine the block diagram below. Write the name of the labeled feature on each of the blanks provided.

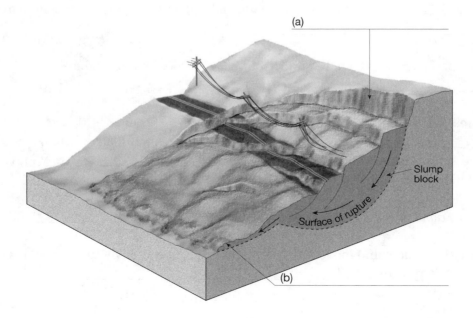

3.   What is mass wasting? _____
   _____
   _____
   _____

4.   What is a slide? _____
   _____
   _____
   _____

5.   What is a fall? _____
   _____
   _____
   _____

6.   What is a flow? _____
   _____
   _____
   _____

ACTIVITIES AND PROBLEMS

1.   Name five factors that play an important role as controls and triggers of mass wasting.
   _____
   _____
   _____
   _____
   _____

2.   Why are mudflows characteristic of semi-arid mountainous regions? _____
   _____
   _____
   _____
   _____

3.   How are debris flows different from earthflows? _____
   _____
   _____
   _____

4.   In general, how are stream valleys produced? _____
_____
_____
_____

5.   How is the process of mass wasting different from the process of erosion? _____
_____
_____
_____
_____

6.   What are the three factors commonly used to classify mass wasting processes? _____
_____
_____
_____

7.   Carefully study the cross section below of a cliff next to an ocean. Predict what may happen
      to the building, and explain how and why this could occur. _____
_____
_____
_____

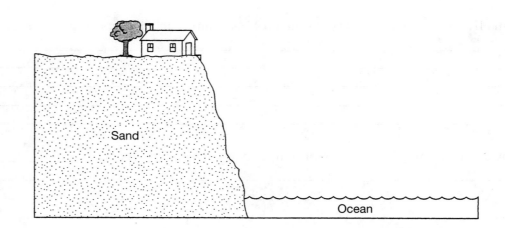

8. Carefully study the cross section below. Predict what may happen to the building, and explain how and why this could occur. _____
_____
_____
_____

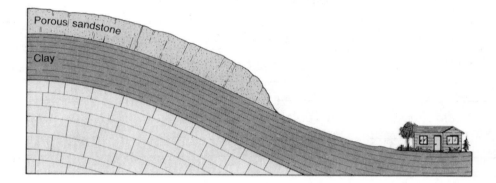

REVIEW EXAM

1. Study the mass wastage examples below. On the line provided below each illustration, name the form of mass wasting that is illustrated (earthflow, slump, rockslide, or debris flow).

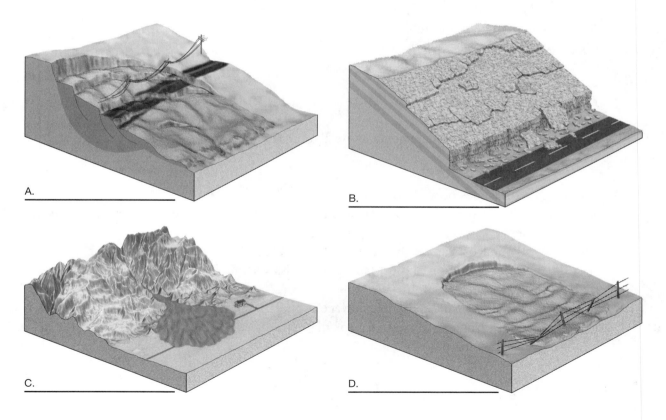

A. _____

B. _____

C. _____

D. _____

2.   What force is the chief cause of mass wasting? _____

3.   What two main effects does water have on mass wasting? _____
_____
_____
_____
_____

4.   Where does permafrost occur in relation to conditions of annual temperature change?
_____
_____
_____

5.   What causes creep to occur? _____
_____
_____
_____
_____

6.   Circle the letter of the most correct answer. Some visible effects of creep are

     a.   tilted fences.
     b.   displaced retaining walls.
     c.   nearly complete lack of vegetation.
     d.   many small channels and gullies.
     e.   a and b.
     f.   all of the above.

7.   For each of the following statements, write T for *true* and f for *false* on the space provided to indicate its validity.

     (a) _____ Landslides occur only on land.
     (b) _____ Mass wasting is a slow process that occurs at speeds of several meters per day or less.
     (c) _____ Permafrost can reach thicknesses as great as 500 meters below the land surface.
     (d) _____ In the evolution of most landforms, mass wasting is the step that follows weathering.
     (e) _____ About 10% of the Earth's surface is underlain by permafrost.
     (f) _____ One advantage of permafrost is that it provides a very firm and durable base on which homes can be built.

CHAPTER 10: RUNNING WATER

LEARNING OBJECTIVES

After reading and studying this chapter, you should be able to

1.   Name and describe processes involved in the hydrologic cycle.

2.   List several factors that control infiltration capacity.

3.   Discuss the importance of velocity in determining the ability of a stream to erode, transport, and deposit sediment.

4.   List four factors that control the velocity of a stream.

5.   Describe the changes that typically occur downstream in velocity, discharge, gradient, channel width, and channel depth.

6.   Discuss the significance of base level and describe the adjustments made by streams when base level is changed.

7.   List three ways that a stream erodes its channel.

8.   Discuss the ways that a stream transports its load.

9.   Distinguish between capacity and competency.

10. Describe the formation of a natural levee and discuss its relationship to the formation of back swamps and yazoo tributaries.

11. Contrast channel deposits and floodplain deposits.

12. Contrast deltas and alluvial fans.

13. Contrast the features and processes associated with a narrow valley and a wide valley.

14. Sketch four different drainage patterns and briefly describe the origin of each.

15. Discuss the concept of headward erosion and its role in stream piracy.

16. Know the two main natural causes of floods and three kinds of flood control measures.

17. Explain what causes widespread floods to occur so frequently on the Red River of the north central United States.

18. Define and understand the key terms listed at the end of the chapter.

VOCABULARY REVIEW

1.   The unending circulation of the Earth's water supply is called the

_____.

2.   Water that soaks into the ground takes the path called (a)_____.
     Water that flows across the surface of the ground is called (b)_____.

3. Some of the water at the Earth's surface is released back into the atmosphere by the process of evaporation. Also, some water that soaks into the ground is absorbed by plants, which later release the water to the atmosphere by the process called
(a)_____. The combined effect of both of these processes is called (b)_____.

4. Water running off the land initially flows in broad, thin sheets. This is called
(a)_____. Next, as currents develop, tiny channels called
(b)_____ form and carry the water to a larger channelized flow of water called a(n) (c)_____.

5. Sheet flow begins when soil becomes saturated and is said to have reached its
_____.

6. The slope of a stream channel is also called the _____.

7. The amount of water flowing past a certain point in a stream over a given unit of time is called _____.

8. There are two ways in which water flows. Water flow in straight-line paths that are parallel to the channel is called (a)_____. Water flow in swirling, erratic paths is called (b)_____.

9. The source area of a stream is called its (a)_____ or
(b)_____. The point downstream where the stream ends, as it enters a larger body of water, is called the (c)_____.

10. The cross-sectional view of a stream, from its source area to the point downstream where it enters a larger body of water, is called a(n) _____.

11. The downward limit to which a stream can erode is called (a)_____.
When such a limit is reached locally (e.g., adjacent to a lake or stream), it is called a(n)
(b)_____. However, sea level is referred to as
(c)_____.

12. A(n) _____ has the correct slope and other channel characteristics necessary to maintain just the velocity required to transport the material supplied to it.

13. Rounded depressions on some riverbeds, created by the abrasive action of swirling particles in fast-moving eddies, are called _____.

14. Streams transport material in three ways. Material transported in solution is called the (a)_____, material transported in suspension is called the (b)_____, and material transported along the channel bottom is called the (c)_____.

15. The speed at which a particle falls through a still liquid is called _____.

16. Raised strips of sediment along the edges of a stream, parallel to the stream, are called (a)_____. Marshes behind these features are called (b)_____, and streams that flow behind such features (parallel to the main stream) are called (c) _____.

17. A(n) _____ is that part of a stream valley that is inundated during floods.

18. _____ streams have an interwoven network of channels.

19. Sediments that bounce or skip along the streambed are said to be moving by _____ _____.

20. The maximum load of solid particles that a stream can transport is called its (a)_____; however, a measure of the maximum size of solid particles that a stream can transport is called its (b)_____.

21. (a)_____ is the process by which solid particles of various sizes are separated from one another during stream transport. The deposit of sediment that results from this process is called (b)_____.

22. Channel deposits are commonly composed of sand and gravel and are called _____.

23. (a)_____ states that every river consists of a main trunk and a variety of branches that form an interrelated system of valleys communicating with one another. The entire land area that contributes water to such a stream system is called a(n) (b)_____, and an imaginary line that separates one of these land areas from another is called a(n) (c)_____.

24. (a)_____ develop where a high-gradient stream leaves a narrow valley in mountainous terrain and abruptly enters onto a broad flat plain or valley floor. In contrast, (b)_____ may have a similar shape, but they form where a stream enters an ocean or lake. Shifting stream channels may also form on these latter features, and such stream channels are called (c)_____.

25. Sweeping bends in sinuous streams that flow in floodplains are called (a)_____. The steep-walled banks on the outside edges of such bends are called (b)_____, and the sedimentary deposits that form on the inside edges of such bends are called (c)_____. When the river cuts through the neck of land in such a bend, the shorter channel is called a(n) (d)_____. The abandoned bend is called a(n) (e)_____ which eventually fills with sediment to form a(n) (f)_____.

26. A steep-walled river valley cut through a ridge or mountain is called a(n) (a)_____, so long as a stream flows through it. If no stream flows through such a valley, then the valley is called a(n) (b)_____.

27. _____ is the process by which a stream can lengthen its course by extending the head of its valley upslope.

28. _____ occurs when the drainage of one stream is diverted into another stream.

29. When a new floodplain is established below an older one, the remnants of the old floodplain are flat surfaces called _____.

30. An undulating plain nearly at base level and caused by the effects of downcutting, sheet flow, and mass wastage, is called a(n) _____.

31. Uplifting the land area on which a meandering mature stream is developed would cause the stream to abandon lateral erosion and revert to downcutting. Streams of this type are said to be (a)_____, and the meanders are known as (b)_____.

APPLYING WHAT YOU HAVE LEARNED

1. What is evapotranspiration? _____
   _____
   _____

2. What is a floodplain? _____
   _____
   _____
   _____
   _____

3. What is a yazoo tributary? _____

    _____

    _____

    _____

4. What is a drainage basin? _____

    _____

    _____

    _____

    _____

5. What is the difference between channel deposits and floodplain deposits? _____

    _____

    _____

    _____

    _____

6. On the spaces provided below each illustration, name the type of drainage pattern that is
    illustrated.

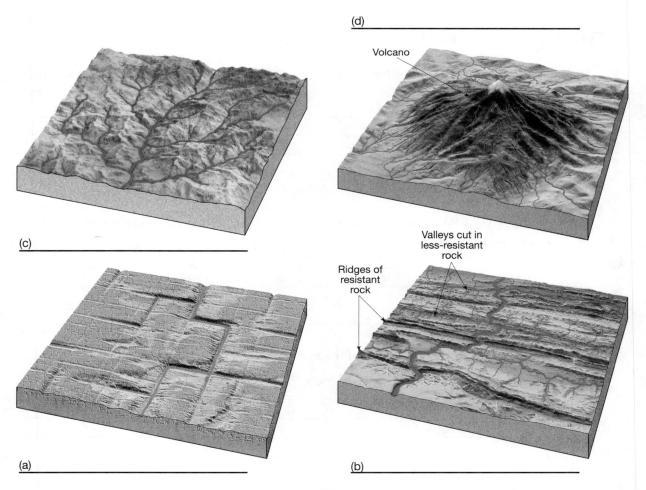

(d) _____

Volcano

(c) _____

Valleys cut in
less-resistant
rock

Ridges of
resistant
rock

(a) _____   (b) _____

91

7. On each of the blanks provided below, fill in the name of the labeled feature.

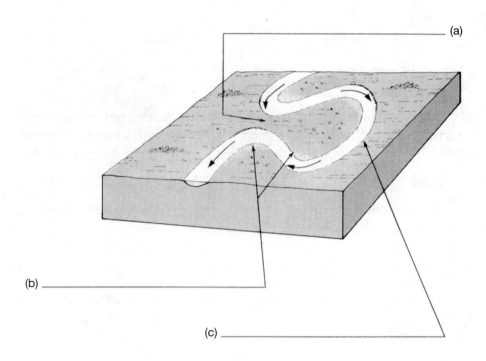

(a) _____

(b) _____

(c) _____

ACTIVITIES AND PROBLEMS

1. If the process of meander cutoff has the effect of shortening a stream, then can it be said that all meandering streams get shorter through time? (a)_____ Why or why not? (b)_____

    _____

    _____

    _____

    _____

2. Describe the process that causes meanders on a river to move laterally, even though the width of the river remains the same. _____

    _____

    _____

    _____

    _____

3. Give two reasons why the Red River of the north central United States floods so frequently.

    _____

    _____

    _____

    _____

4.   Briefly describe how water moves through the water cycle. _____
        _____
        _____
        _____
        _____

5.   Consider a stream that is flowing at a velocity of 2 meters per second through a channel that
        is 10 meters wide and 1 meter deep. What is the discharge of the stream?
        (a)_____ What are two different changes that this stream could
        undergo in order to double its discharge? (b)_____
        _____
        _____

REVIEW EXAM

1.   List at least four factors that control infiltration capacity. _____
        _____
        _____
        _____
        _____

2.   The largest river in North America is the (a)_____ River, but the
        world's largest river is the (b)_____ River.

3.   Circle the letter of the most correct answer. The main factor that determines the size of a
        stream is

        a.   slope.
        b.   valley width.
        c.   discharge.
        d.   load.
        e.   velocity.

4.   Circle the letter of the most correct answer. Of all the water in Earth's hydrosphere, less than
        1% is found in

        a.   oceans.
        b.   ice sheets and glaciers.
        c.   lakes, streams, subsurface water, and the atmosphere.

5. Circle the letter of the most correct answer. The discharge of a stream is

   a. the amount of sediment deposited by a stream.
   b. the amount of water flowing past a point in a stream in a given unit of time.
   c. the amount of water flowing over the banks of a river in a flood.

6. Circle the letter of the most correct answer. A stream erodes downward, deeper into the ground, until it reaches

   a. solid rock.
   b. bottom of the soil.
   c. base level.
   d. both a and b.

7. Circle the letter of the most correct answer. A deposit that forms in the channel of a stream is a(n)

   a. delta.
   b. alluvial fan.
   c. levee.
   d. bar.

8. Circle the letter of the most correct answer. Drainage basins are separated by

   a. yazoo tributaries.
   b. incised meanders.
   c. base levels.
   d. divides.

9. Circle the letter of the most correct answer. Water enters Earth's atmosphere by

   a. evaporation.
   b. evapotranspiration.
   c. infiltration.
   d. both a and c.

10. Name three ways that streams may erode their channels. _____

    _____

    _____

    _____

11. What happens to each of the following qualities as one proceeds downstream in a river?

velocity (a)_____

gradient (b)_____

discharge (c)_____

channel depth (d)_____

channel width (e)_____

12. Name four factors that determine a stream's velocity.

(a)_____

(b)_____

(c)_____

(d)_____

13. Name and briefly describe three ways that a stream transports its load.

_____

_____

_____

_____

_____

_____

14. Study the illustration below, then describe the steps that led to formation of the sharply bent stream that occurs in the valley between the two labeled gaps. _____

_____

_____

_____

_____

_____

_____

_____

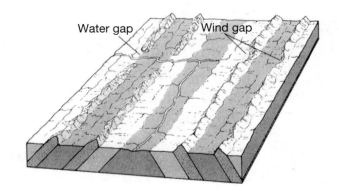

Water gap   Wind gap

15. For each of the following statements, write T for *true* or F for *false* on the space provided to indicate its validity.

(a) _____ The hydrologic cycle is powered chiefly by energy derived from Earth's interior.

(b) _____ A stream's velocity is the primary factor that determines whether the stream's flow is laminar or turbulent.

(c) _____ Heavy rain and snowmelt are the main natural causes of floods.

CHAPTER 11: GROUNDWATER

LEARNING OBJECTIVES

After reading and studying this chapter, you should be able to

1. List the factors that influence the amount of water that soaks into the ground.

2. Describe the distribution of water beneath the land surface in relation to the zone of aeration, zone of saturation, belt of soil moisture, capillary fringe, and water table.

3. Explain the difference between gaining streams and losing streams.

4. Distinguish between porosity and permeability, and between aquifers and aquitards.

5. Describe the movement of groundwater, including the factors controlling the paths of movement, the importance of Darcy's law, and the rate of groundwater movement.

6. Describe or diagram one situation that leads to the formation of a spring.

7. Describe the effect on the water table of pumping from water wells.

8. List two necessary conditions for an artesian system.

9. Name several environmental problems involving groundwater.

10. Understand why groundwater should be treated as a nonrenewable resource.

11. Explain the role of groundwater in the formation of caverns and karst topography.

12. Name two common types of speleothems and distinguish between them.

13. Describe two ways in which sinkholes are created.

14. Explain why land subsidence occurs periodically in California's San Joaquin Valley.

15. Explain why irrigated farming of the High Plains of the U.S. is in jeopardy.

16. Explain how hot springs and geysers form.

17. Define and understand the key terms listed at the end of the chapter.

VOCABULARY REVIEW

1. The percentage of the total volume of rock or sediment that consists of pore spaces is called the _____ of the material.

2. The ability of a material to allow a fluid to move through it is called the _____ of the material.

3. Materials that transmit groundwater are called (a)_____, whereas materials that prevent groundwater movement are called (b)_____.

4.   A (a) _____ stream maintains a flow only because of the influx of water from the groundwater system, but a (b)_____ stream provides water to the groundwater system.

5.   The zone beneath the land surface where molecular attraction holds water as a film on the surface of soil particles is called the (a)_____. Beneath this zone is the (b)_____, where groundwater is lifted against gravity by surface tension in tiny passages between soil particles or grains of sediment. Along with the intermediate belt, these two zones compose the (c)_____.

6.   Water seeping into the ground eventually reaches a zone where all of the open spaces are completely filled with water. This zone is called the (a)_____, and its upper surface is called the (b)_____. Any water in this zone is called (c)_____.

7.   The water table slope is known as the _____.

8.   _____ is the vertical difference in a groundwater system between the point of recharge into the system and the point of discharge from the system.

9.   The upper surface of a localized zone of saturation occurring above the main zone of saturation is called a(n) _____.

10.   A hole that has been dug or drilled into the zone of saturation is called a(n) _____.

11.   A(n) _____ is a natural occurrence of groundwater flowing out onto the surface of the land.

12.   When water is withdrawn from a well, the water table in the vicinity of the well is lowered. This is called (a)_____, and the conical depression of the water table is known as a(n) (b)_____.

13.   A(n) (a)_____ well is one in which groundwater under pressure rises above the level of the aquifer. If the pressure surface is below ground level, then the well is a(n) _____.

14.   A well from which groundwater flows freely without being pumped is called a(n) _____.

15. (a)_____ are intermittent hot springs or fountains in which columns of water are ejected with great force. When water in such features contains dissolved silica, then deposits of (b)_____ or (c)_____ form around them. When such features contain dissolved calcium carbonate, then (d)_____ or (e)_____ is deposited.

16. Groundwater dissolution of limestone may produce spectacular cavities in the bedrock called _____.

17. Dripstone features are collectively called (a)_____. The icicle-like dripstone formations are called (b)_____. Dripstone features that form conical pillars rising from the floor of a cavern are called (c)_____.

18. Landscapes that have been shaped chiefly by the dissolving power of groundwater are said to exhibit _____.

19. Depressions caused by solution of bedrock and collapse of caverns are called (a)_____ or (b)_____.

APPLYING WHAT YOU HAVE LEARNED

1. What is Darcy's law? _____
   _____
   _____
   _____
   _____
   _____

2. Explain in quantitative terms how the hydraulic gradient is obtained. _____
   _____
   _____
   _____
   _____

3. What is a cone of depression? _____
   _____
   _____
   _____
   _____

4. On the cutaway sides of the illustration below, areas shaded dark represent the zone of aeration, and areas that are lightly shaded represent the zone of saturation. All rock types are aquifers except for the labeled aquitard. On the blanks provided, fill in the name of the labeled features.

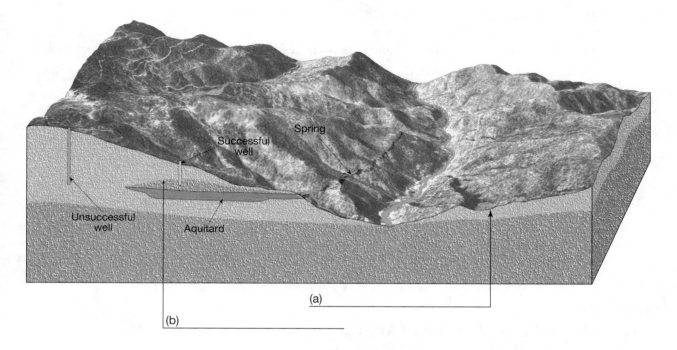

5. On the blank provided beside each cross section below, describe what type of stream is represented. Arrows indicate direction of groundwater flow.

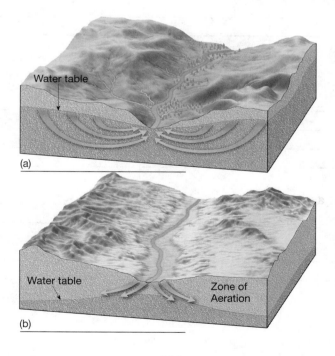

6. What are hot springs, and how do they form? _____
_____
_____

ACTIVITIES AND PROBLEMS

1. Study the photograph below. What is object A called? (a)_____.
   How did object A form? (b)_____
   _____
   _____
   _____

   What is object B called? (c)_____.
   How did object B form? (d)_____
   _____
   _____
   _____

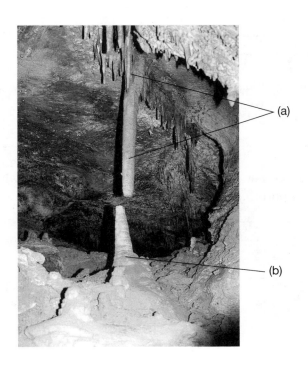

2.   Why are city water systems essentially artificial artesian systems? _____
_____
_____
_____
_____

3.   What are two conditions required to form an artesian system? _____
_____
_____
_____

4.   Why is subsidence sometimes associated with withdrawal of groundwater? _____
_____
_____
_____
_____

5.   Name two ways that saltwater contamination of groundwater under islands like Long Island
can be corrected. _____
_____
_____

6.   How are groundwater contamination problems magnified in extremely permeable aquifers
such as coarse gravel or cavernous limestone? _____
_____
_____

7.   On the space provided beside each material listed below, indicate whether the material is
porous, permeable, or porous and permeable.
sand   (a)_____
Swiss cheese   (b)_____
Styrofoam   (c)_____
a sponge   (d)_____

102

REVIEW EXAM

1. Examine the cross section below, which shows the extent of three wells in relation to the water table. What will happen to the water supply in wells A and C as the wells operate?

   (a)_____ Explain your reasoning.

   (b)_____

   _____

   _____

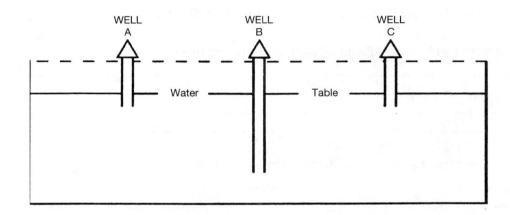

2. List the four main factors that determine the amount of water that soaks into the ground dur-ing rainstorms. _____

   _____

   _____

   _____

3. On the space provided beside each material listed below, indicate whether the material would most likely be an aquifer or an aquitard.

   sandstone   (a)_____

   limestone   (b)_____

   shale   (c)_____

   clay   (d)_____

   gravel   (e)_____

4. Circle the letter of the most correct answer. The energy most responsible for groundwater movement is provided by

   a.   wind.
   b.   gravity.
   c.   stream flow pulling the water along.
   d.   the sun.

5.   Circle the letter of the most correct answer. The water level in a well coincides with

        a.   The upper boundary of the saturated zone.
        b.   the capillary fringe.
        c.   the upper boundary of the zone of aeration.
        d.   the water table.
        e.   both a and d.
        f.   both b and c.

6.   What causes land subsidence to occur in California's San Joaquin Valley? (a)_____
        _____
        What steps can humans take to slow or even prevent this subsidence? (b)_____
        _____
        _____
        _____
        _____

7.   Why is irrigated farming of the High Plains of the U.S. in jeopardy? _____
        _____
        _____
        _____

8.   Study the illustrations below of cross sections of rocks X and Y. Which rock type would
        make the better aquifer? (a)_____ Why? (b)_____
        _____
        _____
        _____
        _____

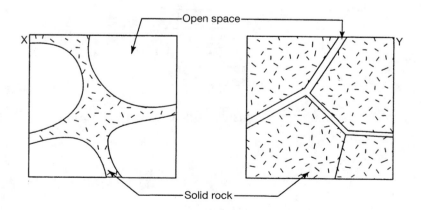

9. For each of the following statements, write T for *true* or F for *false* in the space provided to indicate its validity.

(a) _____ Formation of dripstone always occurs in the zone of aeration.

(b) _____ Groundwater tends to move toward areas of reduced pressure from areas of high pressure.

(c) _____ Artesian wells do not require pumping.

(d) _____ The primary erosional work carried out by groundwater is that of physical abrasion.

(e) _____ The Ogallala Aquifer provides unlimited supplies of water to farms of the High Plains region of the U.S.

(f) _____ The land subsidence in California's San Joaquin Valley has ended, and is no longer a problem for humans to worry about.

CHAPTER 12: GLACIERS AND GLACIATION

LEARNING OBJECTIVES

After reading and studying this chapter, you should be able to

1. Compare the extent and distribution of glaciers today and during the Ice Age.

2. Outline the stages in the formation of glacial ice.

3. Contrast plastic flow and basal slip.

4. Explain how the advance, retreat, or stasis of a glacier relates to the budget of the glacier.

5. List and briefly describe the erosional features associated with alpine glaciers.

6. Distinguish between till and stratified drift.

7. List and distinguish among the basic types of moraines.

8. Describe how drumlins and roches moutonnées may be used to determine the direction of glacial movement.

9. Sketch and/or describe the relative positions of the following features: terminal moraine, outwash plain, ground moraine, kettle holes, recessional moraine.

10. Distinguish between kames and eskers.

11. Briefly discuss the development of glacial theory as an example of applying the principle of uniformitarianism.

12. List at least three indirect effects of Pleistocene glaciers.

13. Briefly discuss how plate tectonics and orbital variations might explain the ice ages.

14. Describe what effects Ice Age glaciers had on organisms, streams, and Earth's crust.

15. Explain what evidence of climate change is recorded in glacial ice.

16. Explain what glacial surges are and why scientists study them.

17. Define and understand the key terms listed at the end of the chapter.

VOCABULARY REVIEW

1. A(n) _____ is a thick mass of ice that originates on land from the accumulation, compaction, and recrystallization of snow.

2. Glaciers that are streams of ice occupying former river valleys are called
   (a)_____ or (b)_____.
   (c)_____ glaciers form where several of these glaciers merge at the base of a mountain.

3. _____ are enormous masses of ice that flow outward in all directions from one or more centers.

4. Glaciers may also flow from the land into the sea. Relatively flat masses of floating ice attached to the land are called _____.

5. _____ resemble ice sheets, but they are smaller, commonly covering uplands and plateaus.

6. Tongues of ice that extend outward from the margins of ice sheets are called _____.

7. Granular recrystallized snow is called _____.

8. Ice flow is basically of two types. (a)_____ occurs when the glacier slides over the ground, and (b)_____ occurs when there is movement within the ice.

9. The upper brittle zone of a glacier is called the (a)_____. Large open cracks that form in this zone are called (b)_____.

10. The advance of some glaciers is characterized by periods of extremely rapid movements called _____.

11. Snow accumulation and formation of glacial ice occur in a region called the (a)_____. The downhill limits of this region are defined by the (b)_____.

12. The loss, or wastage, of glacial ice is termed _____.

13. The part of the glacier that occurs below snowline is the _____.

14. Large pieces of ice may break off the front of glaciers in a process called _____.

15. _____ is the process whereby glaciers loosen and lift blocks of rock and incorporate them into the ice.

16. Abrasion by glaciers may create long scratches and grooves on bedrock surfaces, called _____. Rock pulverized during glacial abrasion is called rock flour.

17. The _____ is the balance, or lack of balance, between accumulation at the upper end of the glacier and ablation at the lower end of the glacier.

18. As glaciers flow around sharp curves, they may remove spurs of land that formerly extended into the valley. The results of this process are triangular-shaped cliffs called (a)_____. Valleys feeding into the main valley glacier can also be truncated by the main valley glacier and are then left standing high above the main glaciated trough. Such valleys are called (b)_____.

19. Series of water-filled bedrock depressions on the floor of a glaciated valley are called
_____.

20. Bowl-shaped depressions at the heads of glaciated valleys are called (a)_____. If small lakes are developed in such features, then they are called (b)_____.

21. A(n) _____ is a mountain pass or gap created when the common headway between two glaciers is eliminated by glacial erosion and frost action.

22. _____ are deep, steep-sided glacial troughs that have been flooded by the ocean.

23. Long knife-edged ridges formed by glacial erosion of divides between adjacent glaciated valleys are called (a)_____. Pyramidal mountain peaks associated with such features are called (b)_____.

24. _____ are knobs of bedrock formed when glacial abrasion smooths slopes facing the oncoming ice and steepens slopes facing the opposite direction.

25. (a)_____ is the name applied in general to any sediment of glacial origin. If such sediment has been layered and sorted by water, then it is called (b)_____. If such sediment is unstratified (deposited directly from the glacier), then it is called (c)_____.

26. Boulders in till, or isolated from other sedimentary particles, are called _____ if they are different from the bedrock beneath them.

27. A ridge of till that marks the farthest advance of an alpine or continental glacier is called a(n) (a)_____ or a(n) (b)_____. If, however, such moraines mark positions where the ice front occasionally stabilized during retreat, then they are called (c)_____.

28. A relatively thin layer of till blanketing the region where a glacier once existed is called a(n) _____.

29. Ridges of till that form along the sides of a valley glacier are called
_____.

30. _____ are ridges of till that form when two alpine glaciers merge to form a single ice stream.

31. Smooth, elongated, parallel hills of till are called (a)_____, whereas mounds of stratified drift are called (b)_____.

32. _____ form when drift fills cracks and other depressions in stagnant ice.

33. Terraces composed of stratified drift and that occur along the sides of a glaciated valley are called _____.

34. A sheet-like deposit of stratified drift confined to the floor of a mountain valley is called a(n) (a)_____, but sheet-like deposits of stratified drift that blanket larger regions are called (b)_____. Depressions in such deposits are called (c)_____.

APPLYING WHAT YOU HAVE LEARNED

1. What is a continental ice sheet?

   _____

   _____

   _____

   _____

2. What are glacial erratics?

   _____

   _____

   _____

   _____

3. What are cirques?

   _____

   _____

   _____

   _____

4. What basic difference is there between the terrain of regions eroded by ice sheets and the terrain of regions eroded by alpine glaciers?

   _____

   _____

   _____

   _____

5. Name two examples of modern ice sheets.

_____

_____

6. On the blanks provided below, fill in the name of the labeled feature of Alpine (mountain) glaciation.

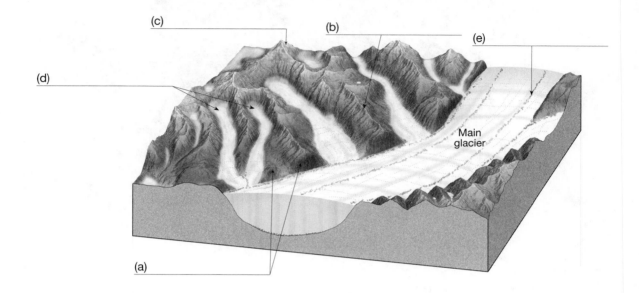

7. What is the difference in shape between a valley eroded chiefly by a river and a valley eroded chiefly by a glacier? (a) _____

_____

Describe how these differences are caused. (b) _____

_____

_____

8. On the blanks provided below, fill in the name of the labeled feature that formed as a product of Alpine glaciation

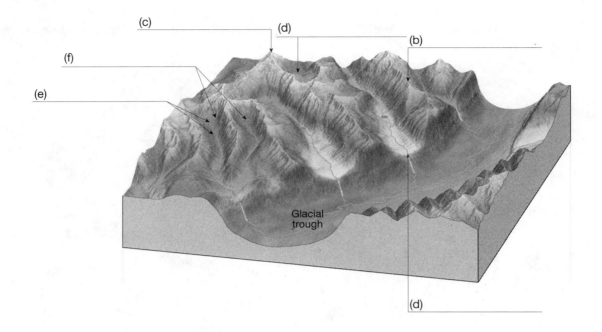

9. What are glacial surges? (a) _____

_____

_____

Why do scientists study glacial surges? (b) _____

_____

_____

_____

10. How was the development of glacial theory an example of applying the principle of uniformitarianism?

_____

_____

_____

_____

_____

_____

11.   On the blanks provided below, fill in the name of the labeled feature.

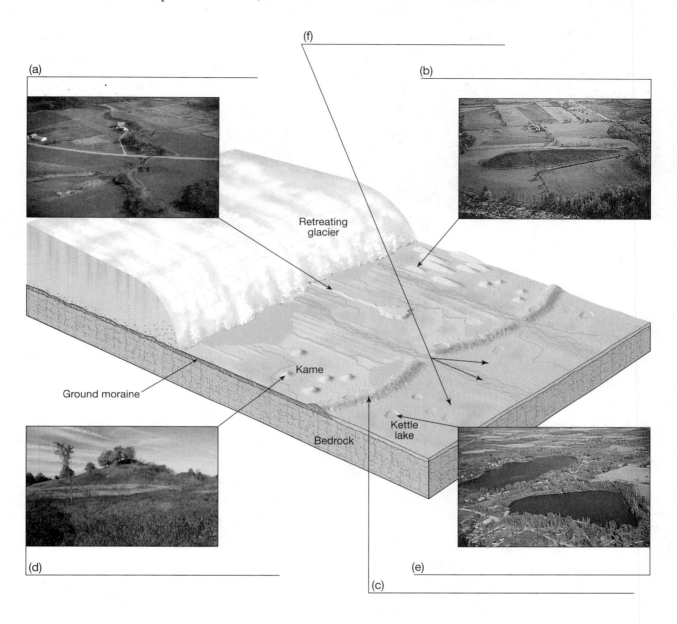

(f) _____

(a) _____

(b) _____

Retreating glacier

Ground moraine

Kame

Kettle lake

Bedrock

(d) _____

(e) _____

(c) _____

ACTIVITIES AND PROBLEMS

1.   Examine the map of glacial striations (black lines) below. At what location (letter) did the glacier originate that made the striations? _____

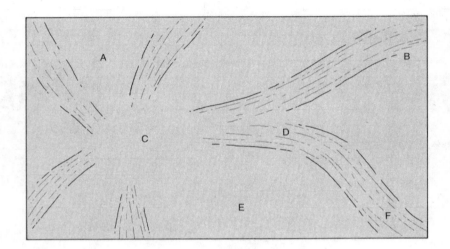

2.   What are recessional moraines?

        _____

        _____

        _____

        _____

3.   What are four factors that control erosion by glacial ice?

        (a) _____

        (b) _____

        (c) _____

        (d) _____

4.   Outline the stages in the formation of glacial ice. _____

        _____

        _____

        _____

        _____

        _____

5.   Briefly discuss the development of glacial theory as an example of applying the principle of uniformitarianism.

        _____

        _____

        _____

        _____

        _____

6.   Study the photograph below. What is the name of the valley from which the waterfall is flowing? (a) _____
How did it form? (b) _____
_____
_____
_____
_____
_____

REVIEW EXAM

1.   List three indirect effects of Pleistocene glaciers.

_____
_____
_____
_____

2.   How could the theory of plate tectonics be used to explain ice ages?

_____
_____
_____
_____
_____
_____

3.   How can orbital variations of the Earth be used to explain the ice ages?

_____
_____
_____
_____
_____
_____

4.   What are the two primary ways that glaciers erode the land?

_____
_____
_____
_____

5.   How do fiords form?

_____
_____
_____
_____
_____

6.   How does the percentage of Earth's land presently covered by glaciers compare to the percentage of land covered during the last Pleistocene glaciation?

_____
_____
_____

7. How does glacial ice provide scientists with information about how Earth's climate has changed?

_____

_____

_____

_____

_____

_____

_____

8. Circle the letter of the most correct answer. If the Antarctic ice sheet were to melt completely, then about how much would sea level rise?

a.   an insignificant amount.
b.   6–10 meters
c.   60–70 meters
d.   200–230 meters
e.   600–700 meters

9. Circle the letter of the most correct answer. Today, glaciers cover what percentage of Earth's surface?

a.   1%
b.   2%
c.   10%
d.   20%

10. Circle the letter of the most correct answer. The boundary between the zone of accumulation and the zone of ablation is

a.   a crevasse.
b.   the glacial budget.
c.   a col.
d.   the snowline.

11. Circle the letter of the most correct answer. The smallest glaciers are

a.   cirques.
b.   valley glaciers.
c.   cols.
d.   ice sheets.

12. Circle the letter of the most correct answer. Earth's last great Ice Age occurred during what interval of geologic time?

    a.   Precambrian
    b.   Mesozoic
    c.   Pleistocene
    d.   Pluvial

13. For each of the following statements, write T for *true* or F for *false* on the space provided to indicate its validity.

    (a) _____ Texas was partly covered by the Laurentide ice sheet when it reached its maximum extent during the Pleistocene Epoch.

    (b) _____ The Great Lakes were scoured by glaciers.

    (c) _____ Calving is the primary means by which the margins of the Antarctic ice sheet lose ice.

CHAPTER 13: DESERTS AND WINDS

LEARNING OBJECTIVES

After reading and studying this chapter you should be able to

1.   List two common misconceptions about deserts.

2.   Describe the distribution and causes of dry lands.

3.   Explain why the boundary between dry and humid climates cannot be defined by a single rainfall figure (set amount of annual precipitation).

4.   Contrast weathering and erosion in arid and humid areas.

5.   Describe the manner in which wind transports sediment.

6.   Describe the formation and migration of sand dunes.

7.   Name and distinguish among the six different basic types of sand dunes.

8.   Briefly describe the origin and distribution of loess.

9.   Describe the features and characteristics associated with each of the stages of evolution of a desert landscape like the Great Basin region of the U.S.

10. Understand how humans can cause desertification.

11. Understand what is meant by "dry" climate.

12. Describe how Australia's Ayers Rock formed.

13. Explain how people have caused the Aral Sea to dry up.

14. Define and understand the key terms listed at the end of the chapter.

VOCABULARY REVIEW

1.   _____ are plants with waxy leaves, stems, or branches that help make them highly tolerant of drought.

2.   Dry land areas are regarded as (a)_____ if they are truly arid and and as (b)_____ if they are semiarid.

3.   The air-pressure belt centered around the equator is called the (a)_____, whereas the low-latitude air-pressure belts just north and south of the equatorial belt are called (b)_____.

4.   Arid regions on the leeward sides of mountains are called _____.

5. The process of lowering the land by blowing away loose material is called (a)_____. Shallow depressions caused by this process are called (b)_____, and stony veneers developed on desert surfaces lowered by this process are called (c)_____.

6. Wind erodes rock mainly by the process of (a)_____. Stones that have undergone shaping by this sandblasting process are called (b)_____. A (c) _____ is a streamlined, wind-sculpted ridge that is oriented parallel to the prevailing wind.

7. Winds commonly transport sand, which is eventually deposited in mounds or ridges called (a)_____. The leeward slope of such a feature is called the (b)_____, and sloping layers within such a feature are called (c)_____.

8. Crescent-shaped dunes with tips pointing downwind are called (a)_____, whereas small, crescent-shaped dunes with tips pointing into the wind are called (b)_____.

9. Linear sand dunes oriented at right angles to the prevailing wind are called (a)_____, whereas scalloped ridges of sand oriented more or less at right angles to the prevailing wind are called (b)_____. Linear sand dunes oriented more or less parallel to the prevailing wind are called (c)_____.

10. A(n) _____ stream is a stream that is dry except for brief periods of flow after rainfalls.

11. _____ is windblown silt that forms compact sheet-like deposits having no internal layering.

12. _____ is characterized by a pattern of intermittent streams that drain into a region but do not drain out of the region.

13. A(n) (a)_____ is a fan-shaped deposit of sediment formed where a stream leaves a narrow channel in a mountainous area and enters a level plain or valley. If several of these deposits coalesce along the foot of a mountain, then such a feature is called a(n) (b)_____.

14. Shallow ephemeral lakes and ponds developed on the floors of basins are called (a)_____. When such bodies of water evaporate, the flat surfaces that remain are called (b)_____.

15. A sloping, eroded bedrock surface at the base of a mountain in a dry region is called a(n) _____.

16. Isolated erosional remnants of bedrock that rise above the surface of sediment-filled basins are called _____.

17. The expansion of desertlike conditions into nondesert areas is called _____ _____.

APPLYING WHAT YOU HAVE LEARNED

1. What are blowouts?

   _____
   _____
   _____
   _____

2. What are ventifacts?

   _____
   _____
   _____
   _____

3. On the blank space provided beside each illustration below, write the name of the type of dunes shown in the illustration.

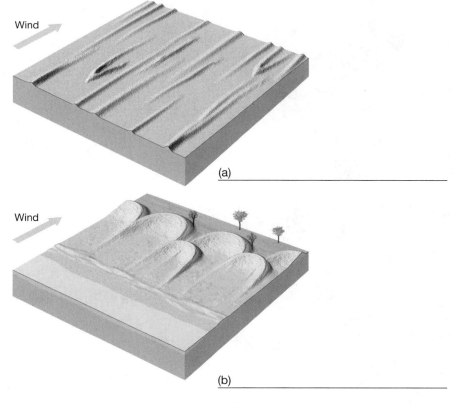

(a) _____

(b) _____

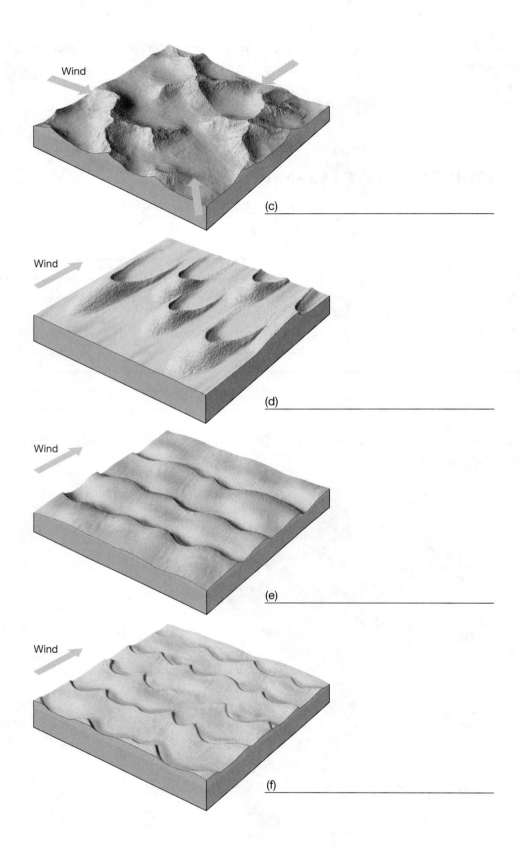

Wind

(c) _____

Wind

(d) _____

Wind

(e) _____

Wind

(f) _____

122

ACTIVITIES AND PROBLEMS

1. What are three causes for dry climates?

   _____
   _____
   _____

2. How does the wind transport sediment?

   _____
   _____
   _____
   _____
   _____
   _____

3. What differences are there between weathering and erosion in humid areas and weathering and erosion in arid areas?

   _____
   _____
   _____
   _____
   _____

4. What are at least three factors that influence the size and form of a sand dune?

   _____
   _____
   _____
   _____

5. Imagine that you are thinking of buying one of the identical homes below, which occur next to two beautiful modem sand dunes. Which home would you choose to buy, if you want to keep it from being covered by sand? (a)_____ Explain (b)_____

   _____
   _____
   _____
   _____

Home A   Home B

6. Imagine that you are thinking of buying one of the identical homes below, which occur next to two beautiful modem sand dunes. Which home would you choose to buy, if you want to keep it from being covered by sand? (a)_____ Explain (b)_____

_____
_____
_____
_____

7. List characteristics of each of the generalized stages below of landscape evolution in a mountainous desert.

Early stage: (a) _____
_____
_____
_____

Middle stage: (b) _____
_____
_____
_____

Late stage: (c) _____
_____
_____
_____

8. How do rainshadow deserts form?

_____
_____
_____
_____
_____

9. What are two common misconceptions about deserts?

_____
_____

124

10. Examine this cross section through the floor of a desert. The cross section is 1 meter wide. What is the name applied to the concentrated layer of rocks on the land surface at the top of this cross section? (a) _____.

How did this layer of rocks form? (b) _____
_____
_____
_____
_____

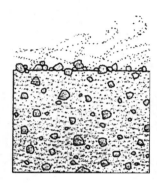

REVIEW EXAM

1. Examine the illustration below of an arid landscape and fill in the blanks with the names of the labeled features.

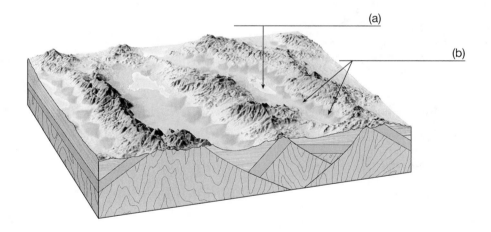

(a)

(b)

Is this an example of an early stage, middle stage, or late stage of landscape evolution in a mountainous desert region (c) _____. Give reasons for your answer. (d) _____
_____
_____

2. What is a "dry" climate?

_____

_____

_____

_____

_____

3. Give two different examples of human-induced activities that contribute to desertification on the margins of existing deserts.

_____

_____

_____

_____

4. How does plate tectonics play a role in forming deserts? _____

_____

_____

_____

_____

_____

_____

5. Why is it that deserts are not defined by a specific, single rainfall figure?

_____

_____

_____

_____

_____

6. Circle the letter of the most correct answer. What percentage of the Earths land surface is dry land?

   a. about 2%
   b. about 10%
   c. about 30%
   d. about 50%

7. Circle the letter of the most correct answer. The word "desert" literally means

   a. deserted.
   b. dry.
   c. the end.
   d. thirst.

8.   Circle the letter of the most correct answer. Wind erosion creates

       a.   ventifacts.
       b.   steppes.
       c.   yardangs.
       d.   both a and b.

9.   Circle the letter of the most correct answer. Sandy wind deposits are called dunes, but silty wind deposits are called

       a.   blowouts.
       b.   loess.
       c.   ventifacts.
       d.   playas.

10.   Circle the letter of the most correct answer. The Aral Sea dried up because
       a.   the climate changed.
       b.   blowing sand filled in the sea.
       c.   mountain waters that supplied the sea were instead diverted for agriculture.

11.   For each of the following statements, write T for *true* or F for *false* on the space provided to indicate its validity.

       (a) _____   A "dry" climate results when yearly precipitation is not as great as the potential loss of water by evaporation.
       (b) _____   The main agent of erosion in deserts is the wind.
       (c) _____   The most common feature of deserts is sand dunes.
       (d) _____   The greatest depth to which a blowout can develop is determined by wind velocity.
       (e) _____   Australia's Ayers Rock is an example of an inselberg.

CHAPTER 14: SHORELINES

LEARNING OBJECTIVES

After reading and studying this chapter, you should be able to

1.   List three factors that control the height, length, and period of a wave.

2.   Describe the circumstances that cause waves to break.

3.   Explain wave refraction and describe its effects along an irregular coastline.

4.   Explain the processes responsible for longshore transport.

5.   Describe and briefly explain the formation of the common shoreline features.

6.   Provide three possible explanations for the formation of barrier islands.

7.   Explain the consequences of artificial structures built for the purposes of modifying shoreline processes.

8.   Explain how increases in atmospheric carbon dioxide relate to global rise in sea level.

9.   Contrast the shoreline erosion problems experienced along the Pacific coast with those occurring along the Atlantic coast.

10.   Define and compare emergent and submergent coasts.

11.   Describe three categories of hurricane destruction.

12.   Briefly explain the cause of tides.

13.   Infer why Louisiana's coastal wetlands are shrinking.

14.   Understand the possible effects of global warming on sea level and the magnitude of destruction by coastal storms.

15.   Define and understand the key terms listed at the end of the chapter.

VOCABULARY REVIEW

1.   The vertical distance between the trough and the crest of a wave is called the
     (a)_____, whereas the horizontal distance separating two successive wave crests is called the (b)_____.

2.   _____is the horizontal distance that wind travels across open water.

3.   As wind blows across a body of water, the height of the waves increases until a critical point is reached where open-water breakers form, called _____ .

4.   Waves in open water move water particles in a circular path and are called
     (a)_____. When a wave breaks, water particles no longer flow in
     a circular path. Rather, the turbulent water advances up the beach slope. These waves are
     called (b)_____.

5.   The turbulent water created by breaking waves is called the _____.

6.   The turbulent sheet of water that washes up the slope of the beach after a breaker collapses is
     called (a)_____. When this water flows back down the beach, it
     is called (b)_____.

7.   The bending of waves is called _____.

8.   Sediment transported in a zigzag pattern along the beach is called (a)_____
     _____. Currents of the surf zone that flow parallel to shore are
     called (b)_____.

9.   The erosional action of the surf acting upon coastal land produces steep slopes called
     (a)_____. As such features retreat, due to continued attack by
     waves, beach-like bedrock surfaces, called (b)_____, may
     develop in front of them.

10. When two caves on opposite sides of a narrow headland unite, a _____
     _____ forms.

11. Isolated remnants of eroded headlands, which have resisted erosion and protrude from
     shallow coastal waters, are called _____.

12. Elongated ridges of sand that project from the land into the mouth of an adjacent bay are
     called _____.

13. A wall built perpendicular to shore for the purpose of trapping sand is called a
     (a)_____, but a wall built parallel to shore to prevent waves from
     eroding the shore areas behind it is called a (b)_____.

14. Low sandy islands that are elongated parallel to shore and separated from the mainland by a
     lagoon are called _____.

15. A(n) _____is a ridge of sand that connects an island to the
     mainland or to another island.

16. A sandbar that completely crosses the mouth of a bay is called a
     _____.

17. _____ are walls constructed in pairs at the entrances to harbors and extending slightly out into the ocean to protect the harbor from wave attack.

18. Walls constructed parallel to shore, but slightly offshore, are called _____ _____.

19. _____ is the process of artificially adding large quantities of sand to a beach.

20. (a)_____ are shorelines created when sea level rises or the land adjacent to the sea subsides. On the other hand, (b)_____ are shorelines created when sea level falls or the land adjacent to the sea is uplifted.

21. Drowned river mouths are called _____.

22. (a)_____ are rhythmic changes in the elevation of the ocean surface at a specific location. When the ocean surface rises, water flows shoreward and is called a(n) (b)_____. When the ocean surface falls, water flows seaward and is called a(n) (c)_____. Portions of the coastline that are affected by these alternating rhythms of water flow are called (d)_____, and the currents associated with the water flow are called (e)_____.

23. _____ exhibit the largest variations between high and low tide because they occur when the sun and moon are aligned and their gravitational effects are added together.

24. _____ exhibit the least variations between high and low tide because they occur when the gravitational effects of the sun and moon are at right angles.

25. A sandy deposit created by tidal currents on the landward or seaward end of a tidal inlet is called a(n) _____.

APPLYING WHAT YOU HAVE LEARNED

1. What is an emergent coast? _____
_____
_____
_____

2. What is flood tide? _____
_____
_____

3. What is beach drift? _____
    _____
    _____

4. What is a tombolo? _____
    _____
    _____

5. On the blanks provided below, write the name of the feature that is labeled.

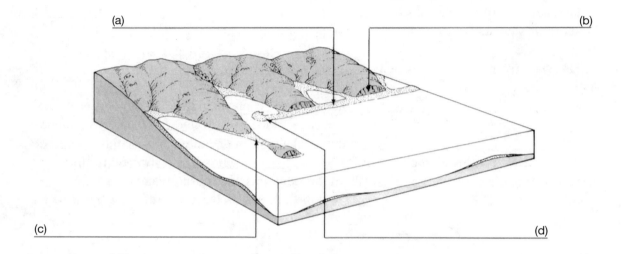

6. Define each of the following:

    Baymouth bar (a) _____
    _____
    _____

    Sea stack (b) _____
    _____
    _____

    Barrier island (c) _____
    _____
    _____

    Groin (d) _____
    _____
    _____

ACTIVITIES AND PROBLEMS

1. What are the basic differences between east-coast shorelines and west-coast shorelines, based on their geologic origins?

   _____
   _____
   _____
   _____
   _____

2. What are two causes of shoreline erosion along the Pacific coast, in addition to sea-level rise?
   (a)_____
   _____
   (b)_____
   _____

3. Why is rising sea level overlooked by many coastal residents as a significant contributor to shoreline erosion problems? _____
   _____
   _____
   _____
   _____

4. Suggest three actions that people should take to protect buildings in coastal areas, in response to shoreline erosion problems. (a)_____
   _____
   (b)_____
   _____
   (c)_____

5. What are the three main factors that determine the height, length, and period of a wave?
   _____
   _____
   _____

6. Describe the process that causes a wave to break.
   _____
   _____
   _____
   _____

7. What is wave refraction? _____
   _____
   _____
   _____

8.   What are the long-term effects of wave refraction upon an irregular coastline?

_____

_____

_____

_____

9.   What are the three main categories of hurricane destruction? _____

_____

_____

REVIEW EXAM

1.   What causes tides? _____

_____

_____

_____

_____

2.   Carefully examine the map below.

What name is given to walls like the wall above that is labeled A?
(a)_____ What name is given to walls like the wall above that
is labeled B? (b)_____ What name is given to walls like the
walls above that are labeled C? (c)_____ In the future, what
will eventually happen to the beach at the following locations?

Location G (d) _____
Location F (e) _____
Location E (f) _____
Location D (g) _____

3. What are three possible ways that barrier islands could form?

(a)_____

_____

(b)_____

_____

(c)_____

_____

4. How is the increase in atmospheric carbon dioxide thought to be causing the global rise in sea level? _____

_____

_____

_____

_____

_____

_____

5. On the blanks provided below, write the name of the feature that is labeled.

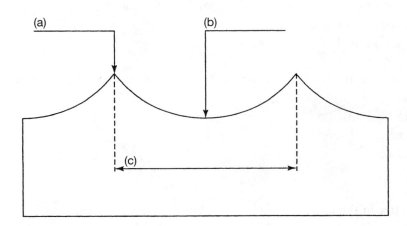

6. Circle the letter of the most correct answer. The most common feature of emergent coastlines is

    a.   wave-cut cliffs.
    b.   sea stacks.
    c.   elevated terraces.
    d.   estuaries.
    e.   tidal flats.

7.   Circle the letter of the most correct answer. The most common feature of submergent coastlines is

   a.   wave-cut cliffs.
   b.   sea stacks.
   c.   elevated terraces.
   d.   estuaries.
   e.   tidal flats.

8.   Circle the letter of the most correct answer. Tides are caused by

   a.   the gravitational attraction exerted upon Earth by the moon and sun.
   b.   Earth's magnetism and wind.
   c.   wind and the shape of coastlines.
   d.   the gravitational attraction exerted upon Earth by the sun.

9.   Circle the leter of the most correct answer. In deep water, a wave

   a.   moves more slowly than in shallow water along the shore.
   b.   causes a circular motion (oscillation) of water particles.
   c.   transports sediment.
   d.   causes a translation of water particles.

10.   Circle the letter of the most correct answer. Another name for a beach is

   a.   river of sand.
   b.   refraction of sand.
   c.   sand wave.
   d.   estuary.

11.   Circle the letter of the most correct answer. During coastal storms,

   a.   waves erode the beach.
   b.   waves build up the beach.
   c.   wave refraction stops.
   d.   longshore currents stop.

12.   For each of the following statements, write T for *true* or F for *false* to indicate its validity.

   (a) _____ Abrasion is probably more intense in the surf zone than in any other environment.
   (b) _____ The average length of a day is increasing because the Earth's rotation is slowing.
   (c) _____ Loss of life from hurricanes Andrew and Hugo was minimal because the affected regions were densely populated with lots of people who were able to help one another.

CHAPTER 15: CRUSTAL DEFORMATION

LEARNING OBJECTIVES

After reading and studying this chapter, you should be able to

1.   Define the forces that deform rock bodies and the basic geologic structures and landforms that result from such deformation.

2.   Describe how confining pressure affects the way rocks behave when deformed.

3.   Know the difference between compressional and tensional stresses.

4.   Describe the two stages of deformation that rocks undergo when great stress is applied to them in an environment of high temperature and high confining pressure.

5.   Briefly describe how mineral composition influences how rock masses will deform.

6.   List three main factors that determine how rocks will behave when subjected to stresses that exceed their own strength.

7.   Name and define the two measurements used by geologists to establish the orientation of rock layers.

8.   Briefly describe how the shapes of large, partly eroded geologic structures are determined.

9.   Name and label the parts of a fold.

10.   Contrast the various types of folds.

11.   Contrast the different types of faults.

12.   Define and distinguish between domes and basins and between horsts and grabens.

13.   Describe the type of stress that produces each type of fault.

14.   Define and distinguish between joints and faults.

15.   Understand what landforms and drainage features are characteristic of regions with active strike-slip faults (like along the San Andreas fault system).

16.   Define and understand the key terms listed at the end of the chapter.

VOCABULARY REVIEW

1.   _____ is the word used to refer to all changes in volume or shape of a rock body.

2.   An applied _____ tends to put stationary objects in motion or change the motion of moving bodies.

3.   Rocks that fracture when deformed are said to undergo (a)_____, whereas rocks that flow when deformed are said to be (b)_____.

4.   (a)_____ is the amount of force acting upon a body of rock to change its shape and/or volume. (b)_____ is the resulting change in shape and/or volume.

5.   Rocks that return to their original shape after the stress is removed are said to deform (a)_____ _____. Deformed rocks that have folded and flowed and remain deformed even after the stress is removed are said to have undergone (b)_____ _____.

6.   Fractures, faults, and folds are three types of _____.

7.   Stress that causes shortening of a rock body is called (a)_____ _____. Stress that elongates rock bodies by pulling them apart is called (b)_____. Stress applied unequally in different directions is called (c)_____.

8.   Sites where bedrock is exposed at the Earth's surface are called _____.

9.   (a)_____ is the direction of the line that is produced by the intersection that an inclined layer of rock makes with an imaginary horizontal surface. (b)_____ is the angle at which the surface is inclined, in a direction perpendicular to the line referred to in the preceding sentence.

10.   _____ are geologic structures formed when flat-lying rocks have been bent into wave-like undulations.

11.   The two sides of a fold are called (a)_____ and the imaginary plane that divides the fold in half as symmetrically as possible is called the (b)_____ _____. The line formed where this imaginary plane intersects a layer within the fold is called the (c)_____ of the fold.

12.   A fold in which the axis is not horizontal is said to be _____.

13.   An arch-like fold (upfold) having the oldest layers in its center is called a(n) (a)_____ _____, whereas a trough-like fold (downfold) having the youngest layers in its center is called a(n) (b)_____.

14.   If the limbs of a fold diverge at the same angle from the axial plane, then the fold is said to be (a)_____, but if the limbs of a fold diverge at different angles from the axial plane, then the fold is said to be (b)_____.

15.   A fold is said to be _____ if one limb is tilted beyond vertical, so that it is upside down.

16.   A fold that has tilted so much that it is "lying on its side" is said to be

_____.

17.   Broad flexures that have just one limb and result from vertical displacement are called

_____.

18.   Broad, blister-like upwarps of rock layers are called (a)_____,
whereas broad, dimple-like downwarps of rock layers are called
(b)_____.

19.   _____ form when exposed ends of resistant inclined strata form
conspicuous ridges that stand out above the more eroded, less resistant, adjacent strata.

20.   (a)_____ are fractures along which displacement has occurred.
When such features are very large, they generally consist of many closely spaced,
interconnected surfaces in a broad belt of deformation known as a(n)
(b)_____.

21.   (a)_____ is a clayey material formed by the pulverizing and
grinding of rocks along fault surfaces. Striations or linear grooves that occur on the fault
surfaces, as a result of the grinding, are called (b)_____.

22.   Faults along which movements are primarily vertical are called
(a)_____. In such faults, the rock surface immediately
above the fault is called the (b)_____, and the rock surface
immediately below the fault is called the (c)_____.

23.   Faults along which movements are primarily horizontal (and parallel to the strike of the fault)
are called _____.

24.   _____ are low cliffs formed where fault displacements
extend to the surface of the land.

25.   An isolated remnant of an eroded thrust sheet is called a(n) _____
_____.

26.   Dip-slip faults are classified as (a)_____ when the hanging
side wall slides down relative to the footwall and are classified as
(b)_____ when the hanging wall slides upward relative to
the footwall. If the latter type of fault is inclined to a nearly horizontal position, then the
faults are instead called (c)_____.

27.   When the slopes of several adjacent normal faults decreases with depth and they join to form
a single, nearly horizontal fault, the result is a _____ fault.

28   Mountains caused by the upward movements of huge blocks of lithosphere along normal faults are called _____.

29.   Uplifted, elongated blocks of crust that are bounded by normal faults are called (a)_____, and elongated valleys between such features are called (b)_____.

30.   A strike-slip fault is said to be _____ if the crustal block on the opposite side of the fault seems to have moved to the right as you look across the fault.

31.   Major strike-slip faults that cut through the lithosphere and accommodate horizontal motions between two major lithospheric plates are called _____.

32.   Rock fractures along which no appreciable movement has occurred are called _____.

APPLYING WHAT YOU HAVE LEARNED

1.   Define each of the following terms:

        Anticline (a)_____
        _____

        Recumbent fold (b)_____
        _____

        Dome (c)_____
        _____

        Plunging fold (d)_____
        _____

2. Carefully study each illustration below to determine the type(s) of geologic structure(s) it contains. On the blanks provided to the right of each illustration, write the name(s) of the structure(s) present in the illustration.

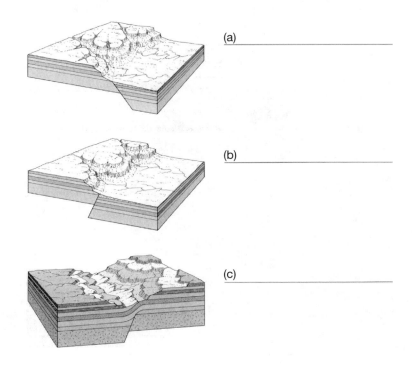

(a) _____

(b) _____

(c) _____

3. On the blanks provided below, fill in the name of the specific type of geologic feature that has been labeled.

(b) _____

(a) _____

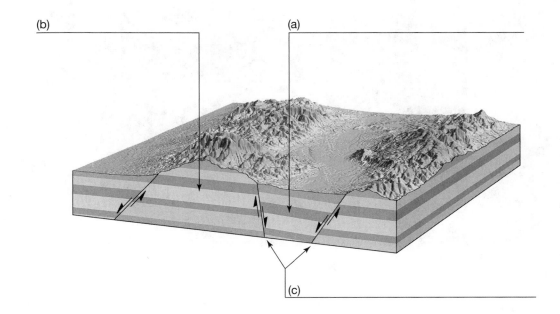

(c) _____

4.   What kind of geologic structure is present in the photograph below?
      (a)_____.
      Explain your answer. (b)_____
      _____
      _____
      _____

5.   What kind of geologic structure is present in the photograph below?
      (a)_____.
      Explain your answer. (b)_____
      _____
      _____
      _____

6. On the blanks provided below, write the name of the specific feature that has been labeled. What kind of fold is this? (d)_____

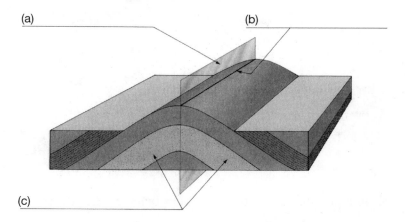

ACTIVITIES AND PROBLEMS

1. Exactly what kind of feature is the San Andreas fault, and exactly what kind of relative motion occurs along that fault? _____
   _____
   _____

2. On the blank provided beside each geologic structure below, indicate whether the structure is caused by compressional stresses, differential stresses, or tensional stresses.

   Syncline (a)_____
   Normal fault (b)_____
   Anticline (c)_____
   Strike-slip fault (d)_____
   Reverse fault (e)_____

3. How are the shapes of large, partly eroded geologic structures determined?
   _____
   _____
   _____

4. How does mineral composition affect how rocks deform? _____
   _____
   _____
   _____
   _____

143

5.　　When do rocks deform? _____
_____
_____

REVIEW EXAM

1.　　What is the difference between a dome and a basin? _____
_____
_____
_____
_____

2.　　How does confining pressure affect the way rocks behave when deformed?
_____
_____
_____
_____

3.　　What kind of geologic structure is present in the photograph below?
(a)_____.
Explain your answer. (b)_____
_____
_____

4.　　What are the three main factors that determine how rocks will behave when subjected to
stresses that exceed their strengths? _____
_____
_____
_____

5. What is strike? _____
    _____
    _____
    _____

6. How can rock bodies be bent into intricate folds without being crushed in the process?
    _____
    _____
    _____

7. Circle the letter of the most correct answer. What landform or drainage feature is *not* characteristic of regions with active strike-slip faults?

    a. sag pond
    b. hogback
    c. linear valley
    d. offset drainage

8. Each of the two block diagrams below portrays a specific kind of geologic structure. Write the name of the structure on the blank provided beside each block diagram.

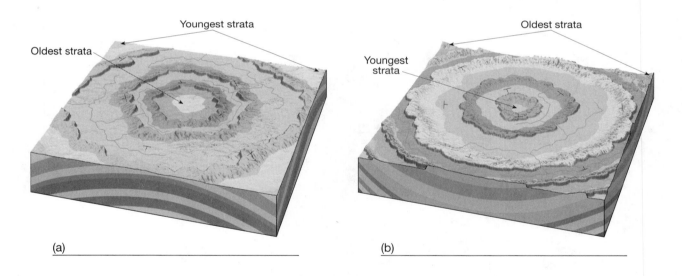

(a) _____ (b) _____

9. For each of the following statements, write T for *true* or F for *false* to indicate its validity.

    (a) _____ Marble benches have been known to sag under their own weight over a period of a hundred years or so.
    (b) _____ In a technical sense, anticlines are simply upfolds.
    (c) _____ The Black Hills of South Dakota are actually eroded remnants of a basin.
    (d) _____ Plunging folds form V-shaped outcrop patterns as they are being eroded.

CHAPTER 16: EARTHQUAKES

LEARNING OBJECTIVES

After reading and studying this chapter, you should be able to

1. Explain how an earthquake is generated and the concept of elastic rebound.

2. Know the difference between surface waves and body waves generated during an earthquake, and how P and S body waves travel through the Earth.

3. Describe how a seismograph works and be able to identify P and S waves on a seismogram.

4. Distinguish between the focus and the epicenter of an earthquake.

5. Contrast foreshocks and aftershocks.

6. Determine the distance between an earthquake and a seismic station, given the P-S interval and a travel-time graph.

7. Describe the world distribution pattern of earthquake activity.

8. Know the difference between earthquake intensity and earthquake magnitude.

9. Discuss the basis of the Modified Mercalli Intensity Scale and briefly describe its limitations.

10. Know how the Richter scale of earthquake magnitude is calculated and used in comparison to the moment magnitude of an earthquake.

11. List the three factors that influence the amount of destruction caused by seismic vibrations.

12. List three types of destruction associated with earthquakes other than the destruction caused directly by seismic vibrations.

13. Briefly describe three different attempts at short-range earthquake prediction.

14. Discuss the possibilities of long-range earthquake prediction.

15. Define and understand the key terms listed at the end of the chapter.

VOCABULARY REVIEW

1. A(n) (a)_____ is the vibration of the Earth produced by rapid release of energy. The source of the energy is called the (b)_____ and the location on the land surface above this source is called the (c)_____.

2. As an earthquake occurs, the elastically deformed rocks along the fault spring back to their original shape as much as possible. This is called _____.

3. Small tremors that precede the main shock in a series of earthquakes are called (a)_____, but small tremors that occur shortly after the main shock are called (b)_____.

4. The study of earthquake waves is called _____.

5. Modern instruments that record seismic waves are called (a)_____ _____, and the actual record of an earthquake is called a(n) (b)_____ _____.

6. Seismic waves that travel only along the outer part of the Earth are referred to by seismologists as either _____ or _____.

7. Body waves travel through the interior of the Earth and are of two types. The (a)_____ _____ push (compress) and pull (expand) rocks in the direction the waves are traveling. The (b)_____ shake rock particles at right angles to the direction the waves are traveling.

8. Earthquakes that occur at depths of less than 40 km within the earth are classified as (a)_____. Those that occur between depths of 70 and 300 km are classified as (b)_____, and those that occur at depths greater that 300 km are classified as (c)_____.

9. The zone in which earthquakes occur beneath an ocean trench is called a(n) (a)_____. Earthquakes in such zones may occur from depths of a few kilometers to depths as great as (b)_____ kilometers.

10. The scale used to measure earthquake intensity on the basis of descriptions of the actual event is called the _____.

11. Earthquake (a)_____ is a measure of the effects of an earthquake at a particular location. In contrast, earthquake (b)_____ _____ is a measure of the strength of an earthquake (amount of energy released).

12. A refined _____ scale is used worldwide to describe earthquake magnitude and is calculated by measuring the amplitude of the largest seismic wave recorded on a seismogram.

13. High-velocity, long-period sea waves that are sometimes generated by submarine earthquakes and often incorrectly referred to as "tidal waves" are more correctly called _____ or _____.

14. Sometimes during earthquakes, stable soil is transformed into a fluid material that is unable to support buildings or other structures. This transformation is called

    _____ .

15. Motions along some faults are smooth, slow, gradual displacements known as
    _____ , which produces few noticeable earthquakes.

16. Some earthquake zones are relatively quiet and have not produced a major earthquake in more than a century. These zones are called _____ .

17. The size of moderate and large earthquakes is expressed as _____ , which is calculated using the average displacement of the fault, the area of the fault surface, and the shear strength of the faulted rock.

APPLYING WHAT YOU HAVE LEARNED

1. List and contrast the three basic types of earthquake waves.
   (a)_____
   _____
   _____
   _____
   _____

   Which type of wave causes the most damage to buildings and other structures?
   (b)_____

2. Carefully study the seismogram below and note the features labeled A through G. On the blanks provided beside each item below, note the letter above that corresponds to that item.

   An S-wave (a)_____   The first S-wave (b)_____
   The first P-wave (c)_____   An L-wave (d)_____
   A surface wave (e)_____

3.   What is a Wadati-Benioff zone? _____

_____

_____

_____

_____

_____

4.   What is earthquake magnitude? _____

_____

_____

_____

_____

5.   What is earthquake intensity? _____

_____

_____

_____

6.   How is the Richter scale of earthquake magnitude calculated and used in comparison to the moment magnitude of an earthquake? _____

_____

_____

_____

_____

_____

_____

_____

_____

ACTIVITIES AND PROBLEMS

1.   Describe why and how an earthquake occurs along a fault. _____

_____

_____

_____

_____

2.   How do seismographs work? _____

_____

_____

_____

_____

3.   How does seismic wave amplitude relate to values of Richter magnitude?

      _____

      _____

      _____

      _____

4.   How much greater are wave amplitudes from an earthquake of Richter magnitude 6.0
      compared with wave amplitudes from an earthquake of Richter magnitude 4.0?

      _____

5.   How much more energy is released during an earthquake of Richter magnitude 6.3 than
      during an earthquake of Richter magnitude 3.3? _____

REVIEW EXAM

1.   Circle the letter of the most correct answer. The best way to describe the strength of a very
      strong earthquake is by determining its

      a.   Modified Mercalli Intensity.
      b.   Richter magnitude.
      c.   moment magnitude.

2.   Circle the letter of the most correct answer. The seismic waves that do the most damage to
      buildings are the

      a.   P-waves.
      b.   S-waves.
      c.   L-waves.
      d.   D-waves.

3.   Circle the letter of the most correct answer. An active fault may experience a moderate earth-
      quake if it

      a.   has seismic gaps.
      b.   has fault creep.
      c.   has buildings constructed on it.
      d.   has liquefaction.

4. Circle the letter of the most correct answer. Normally, the first seismic waves to reach seismographs after an earthquake are

    a.   S-waves.
    b.   P-waves.
    c.   F-waves.
    d.   L-waves.

5. Circle the letter of the most correct answer. Which item below would not produce an earthquake

    a.   cracking of a large mass of rock to release strain energy.
    b.   a nuclear blast.
    c.   fluctuations of Earth's magnetic field.
    d.   the collapse of a building.

6. Circle the letter of the most correct answer. Tsunamis are

    a.   seismic vibrations.
    b.   fires caused by earthquakes.
    c.   areas of Japan that were devastated in the Kobe earthquake.
    d.   seismic sea waves.

7. What are the three main factors that determine how much structural damage to buildings will occur because of an earthquake?

_____
_____
_____
_____
_____

8. Carefully examine the map below, which shows modern, active faults as black lines. Black dots are locations where shallow earthquakes of Richter magnitude 2.5 or less have occurred in the past six months. At which location (A, B, or C) would you predict that a larger earthquake may occur? (a) _____.
Explain your reasoning. (b)_____
_____
_____
_____

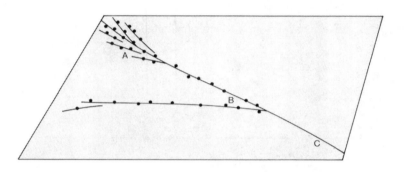

9. How does the nature of the material on which buildings are constructed determine their susceptibility to damage in an earthquake? _____
_____
_____
_____
_____

10. Briefly describe the global pattern of earthquake activity on Earth. _____
_____
_____
_____
_____
_____

11. How are the visible effects of seismic sea waves different in the open ocean than in shallow coastal waters? _____
_____
_____

12.   Study the cross section below. Which of these two identical homes would suffer the most damage if an earthquake occurred along the active fault? (a)_____.
        Explain your answer. (b)_____
        _____
        _____
        _____

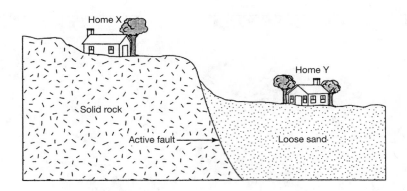

13.   For each of the following statements, write T for *true* or F for *false* on the space provided to indicate its validity.

        (a) _____ Monitoring foreshocks has been used as a means of predicting when a
                    forthcoming major earthquake will occur.
        (b) _____ In the 1906 San Francisco earthquake, much of the damage was caused by fires.
        (c) _____ P-waves are not transmitted through fluids.
        (d) _____ Major earthquakes occur mostly along boundaries between lithospheric plates.
        (e) _____ Earthquakes with a Richter magnitude greater than 9.0 do not occur.
        (f) _____ Damaging earthquakes do not occur east of the Mississippi River.
        (g) _____ Ground-motion amplification of seismic waves occurs when the vibration of
                    earth materials and buildings matches the vibration of the seismic waves.

CHAPTER 17: EARTH'S INTERIOR

LEARNING OBJECTIVES

After reading and studying this chapter, you should be able to

1. List five significant characteristics that relate to the behavior of seismic waves.

2. Briefly describe the contributions of A. Mohorovicic and I. Lehmann to our understanding of Earth's internal structure.

3. Relate the discovery of the shadow zone to the existence of a core.

4. List the four compositional layers of Earth's interior and contrast them in terms of their thickness and chemical makeup.

5. Contrast conduction and convection.

6. Define the five mechanical layers of Earth's interior on the basis of their physical properties and their positions in relation to one another.

7. Explain why meteorites were useful in understanding Earth's interior.

8. Explain why Earth has a magnetic field.

9. Present the seismological and other evidence that indicates that the outer core is liquid.

10. Define seismic tomography and how it is used to study Earth's interior.

11. Define and understand the key terms listed at the end of the chapter.

VOCABULARY REVIEW

1. The boundary between two dissimilar layers within Earth is called a(n) _____.

2. The Earth's interior consists of three compositionl layers. The very thin outer layer is called the (a)_____ and the rocky layer below it is called (b)_____. The (c)_____ is a metallic sphere at the Earth's center.

3. The belt where direct P-waves are absent is called the _____.

4. The boundary between the Earth's crust and mantle is called the (a) _____ _____ or _____. The boundary between Earth's inner core and outer core is the (b) _____ _____.

5. A weak (easily deformed) zone of partly molten rock that is located about 100–700 km within the Earth is called the (a)_____. The brittle zone above it is called the (b)_____.

6. The mantle is divided into the upper mantle (in contact with the crust) and the lower mantle or (a)_____. At the very base of the mantle is located a zone of partially molten rocks called the (b)_____.

7. Temperature inside the Earth gradually rises with increasing depth at a rate known as the _____.

8. The transfer of heat through matter by molecular activity is a process called _____.

9. The most important process in the mantle is the transfer of heat energy by mass movement or circulation. This process is called _____.

APPLYING WHAT YOU HAVE LEARNED

1. What is the P-wave shadow zone? _____
_____
_____
_____
_____

2. What is the asthenosphere? _____
_____
_____
_____
_____

3. What is the lithosphere? _____
_____
_____
_____
_____

4. What is the crust? _____
_____
_____
_____
_____

5. On the blank spaces provided in the illustration below, fill in the names of the layers of Earth's interior that are labeled.

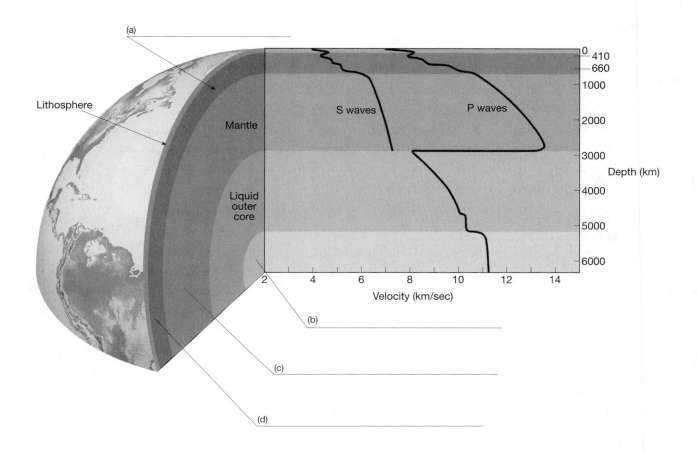

(a) _____

Lithosphere

Mantle

S waves   P waves

Liquid outer core

2   4   6   8   10   12   14
Velocity (km/sec)

0
410
660
1000
2000
3000
4000
5000
6000
Depth (km)

(b) _____

(c) _____

(d) _____

ACTIVITIES AND PROBLEMS

1. What is seismic tomography, and how is it used to study Earth's interior? _____
   _____
   _____
   _____
   _____

2. How is the Earth's magnetic field probably generated? _____
   _____
   _____
   _____

3.	How were meteorites useful in understanding the nature of the Earth's interior? _____
	_____
	_____
	_____
	_____
	_____

4.	List five significant characteristics that relate to the behavior of seismic waves.
	(a)_____
	(b)_____
	(c)_____
	(d)_____
	(e)_____
	_____
	_____
	_____

5.	What evidence is there to suggest that the Earth's outer core is liquid? _____
	_____
	_____
	_____
	_____
	_____
	_____

REVIEW EXAM

1.	What are the three compositional layers of Earth's interior and the thicknesses and generalized chemical makeup of those layers? _____
	_____
	_____
	_____
	_____
	_____
	_____
	_____

2.	What contribution did A. Mohorovicic make to our understanding of the Earth's interior?
	_____
	_____
	_____
	_____

3. How did the discovery of the shadow zone relate to the existence of a core within the Earth?

    _____
    _____
    _____
    _____
    _____

4. Carefully examine the illustration below, which shows how P-waves are transmitted through the Earth. Also note the letters A, B, and C.

    Which level indicates the P-wave shadow zone? (a) _____
    Which level indicates the zone where no direct P-waves are received? (b) _____
    Which level indicates the zone where direct P-waves are received?(c) _____

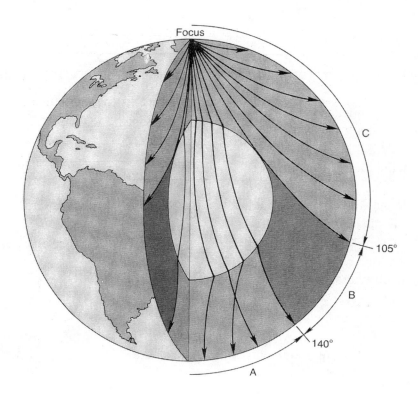

5. Circle the letter of the most correct answer. Inge Lehmann studied the Lehmann discontinuity, which is the boundary between Earth's

    a. crust and mantle.
    b. lithosphere and asthenosphere.
    c. mantle and core.
    d. inner core and outer core.

6.   Circle the letter of the most correct answer. Which layer below is not a compositional layer of Earth's interior?

   a.   core
   b.   asthenosphere
   c.   crust
   d.   mantle

7.   Circle the letter of the most correct answer. The center of Earth is about how far beneath its surface?

   a.   4,000 kilometers
   b.   5,000 miles
   c.   400 kilometers
   d.   500 miles
   e.   1,000 miles

8.   Circle the letter of the most correct answer. Earth's mesosphere is located above its

   a.   lower mantle.
   b.   upper mantle.
   c.   outer core.
   d.   inner core.

9.   Circle the letter of the most correct answer. Earth's compositional layer that is composed mostly of silicon and oxygen is the

   a.   inner core.
   b.   asthenosphere.
   c.   mantle.
   d.   crust.

10. For each of the following statements, write T for *true* or F for *false* on the space provided to indicate its validity.

   (a) _____ Rocks from Earth's mantle occur so deep within the Earth that they have never been observed by humans.
   (b) _____ The Earth's core is probably composed of peridotite.
   (c) _____ Earth's lithosphere rests on the asthenosphere.
   (d) _____ Earth's inner core is liquid.

CHAPTER 18: THE OCEAN FLOOR

LEARNING OBJECTIVES

After reading and studying this chapter, you should be able to

1. Explain what an echo sounder is and how it works.

2. List and describe the three main parts of a passive continental margin.

3. Describe what turbidity currents are and their relationship to the formation of submarine canyons.

4. List and describe the three main features associated with deep-ocean basins.

5. Summarize Darwin's theory on the formation of atolls.

6. Name and give examples of the three broad categories of seafloor sediments.

7. Describe the submarine features of active continental margins and passive continental margins.

8. Describe the relationship among mid-ocean ridges, deep-ocean trenches, and seafloor spreading.

9. Describe the formation of an ophiolite complex.

10. Explain the elevated position of ocean ridges and briefly contrast the Mid-Atlantic Ridge and the East Pacific Rise.

11. Define and understand the key terms listed at the end of the chapter.

VOCABULARY REVIEW

1. An instrument used to do electronic depth-sounding is called a(n) _____.

2. Deep, steep-sided valleys originating on the continental slope are known as (a)_____ _____. Sediment-laden waters that flow through such valleys are called (b)_____, and the deposits that settle out of such flows are called (c)_____. Features that form at the lower end of these submarine valleys resemble alluvial fans and are called (d)_____ _____. They coalesce to produce a continuous apron of sediment at the base of the continental slope that is called the (e)_____.

3. The characteristic of a sedimentary deposit in which sediment grain size decreases from bottom to top is known as _____.

4.   The portion of the seafloor adjacent to a continent is called a(n) (a)_____
     _____. A passive continental margin has three parts. The gently sloping, sub-
     merged surface extending outward from shore is called the (b)_____
     _____. At its seaward margin is a zone that has a steep gradient called the
     (c)_____, which merges into a more gradual seaward incline
     known as the (d)_____.

5.   Between the continental margin and the oceanic ridge system lies the (a)_____
     _____. The extensive flat areas sometimes found in this region are called
     (b)_____.

6.   Long, relatively narrow troughs that form along active continental margins in the deepest
     parts of the oceans are called _____.

7.   Dotting the ocean floors are isolated volcanic peaks called (a)_____.
     Some of these volcanic peaks grow above sea level, are eroded into flat-topped fea-
     tures, and then sink to great depths. Such flat-topped volcanic structures on the ocean
     floor are called (b)_____.

8.   A(n) _____ is a ring of coral reef surrounding a central lagoon that
     formed on the flanks of a sinking volcanic island.

9.   Sea-floor sediments are of three types. (a)_____ consist chiefly
     of mineral grains weathered from continental rocks and transported to the ocean. (b)_____
     _____ consist of shells and skeletons of marine plants and animals. (c)_____
     _____ consist of mineral grains that crystallized directly from sea-
     water via various chemical reactions.

10. Economically important, rounded, blackish lumps of minerals that form very slowly on the
     floors of ocean basins are called _____.

11. Submarine mountains that are the sites of sea-floor spreading are called (a)_____
     _____. The narrow zones at the crest are called (b)_____
     _____. Along some parts of these zones are deep down-faulted valleys called
     (c) _____.

12. The upper layer of ocean-floor rock is (a) _____ basalt. This basalt is
     underlain by a layer of numerous interconnecting dikes called (b)_____
     _____. The lowest layer of the ocean-floor rock sequence is made up of gabbro. This
     three-layer sequence of rocks is called a(n) (c)_____.

13. A chaotic accumulation of deformed sediments and scraps of oceanic crust located along an
     active continental margin is called a(n) _____.

14. _____ are submarine accumulations of the secreted living tissues and skeletal remains of corals and algae.

APPLYING WHAT YOU HAVE LEARNED

1. What is a turbidite? _____
_____
_____
_____

2. What is an abyssal plain? _____
_____
_____
_____

3. What are seamounts? _____
_____
_____
_____

4. What is terrigenous sediment? _____
_____
_____
_____

5. How are seafloor features of an active continental margin different from those of a passive continental margin? _____
_____
_____
_____
_____
_____
_____

ACTIVITIES AND PROBLEMS

1. How are turbidity currents related to the formation of submarine canyons?
_____
_____
_____
_____

2. List and briefly describe the three main features associated with deep-ocean basins.

_____
_____
_____
_____
_____
_____
_____

3. List and describe the three subdivisions of the continental margin.
(a)_____
_____
(b)_____
_____
(c)_____
_____
_____

4. Why is there essentially no calcareous (CaC03) ooze accumulating in the deep-ocean basins?

_____
_____

REVIEW EXAM

1. Summarize Darwin's theory on the formation of atolls. _____
_____
_____
_____
_____
_____

2. List and describe the three broad categories of seafloor sediments.
(a)_____
_____
(b)_____
_____
(c)_____
_____

3. How does an ophiolite complex form? _____
_____
_____
_____
_____
_____
_____

4. Explain the elevated positions of ocean ridges and the differences in relative height between the Mid-Atlantic Ridge and the East Pacific Rise. _____
_____
_____
_____
_____
_____
_____

5. List the steps in J. Tuzo Wilson's hypothesis of how rifting occurs when a thick segment of lithosphere remains stationary over a hot spot for an extended period of time.
_____
_____
_____
_____
_____
_____

6.	The illustration below is a topographic profile of a passive continental margin. Fill in the blanks with the correct name of the feature labeled

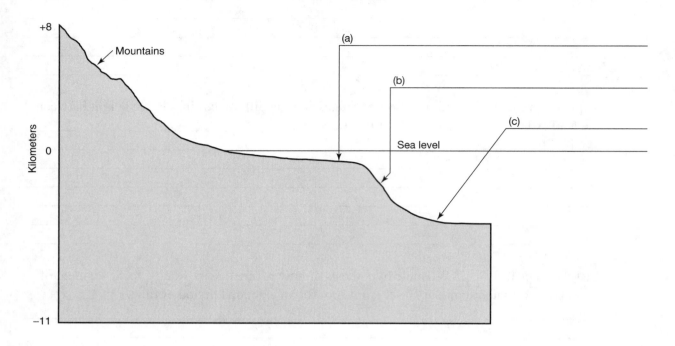

7.	Circle the letter of the most correct answer. Most of the geologic activity associated with a mid-ocean ridge occurs along a narrow region on the ridge called the

	a.	trench.
	b.	atoll.
	c.	rift zone.
	d.	accretionary wedge.

8.	Circle the letter of the most correct answer. Halite crystallized from sea water is an example of

	a.	hydrogenous sediment.
	b.	terrigenous sediment.
	c.	biogenous sediment.

9.	Circle the letter of the most correct answer. An accretionary wedge forms

	a.	where skeletons of coral and algae accumulate.
	b.	where seafloor spreading occurs to form new ocean crust.
	c.	where sediment is scraped from a descending oceanic plate and plastered against the continent along an active margin.
	d.	along the mid-ocean ridge.

10. For each of the following statements, write T for *true* or F for *false* on the blank provided to indicate its validity.

(a) _____ The extreme weight of coral reefs is probably the main reason why atolls sink.

(b) _____ In the geologic future, the Mediterranean Sea will become more narrow.

(c) _____ Calcareous ooze is a type of biogenous sediment.

(d) _____ Manganese nodules are a type of terrigenous sediment.

CHAPTER 19: PLATE TECTONICS

LEARNING OBJECTIVES

After reading and studying this chapter, you should be able to

1. List the evidence that led Wegener to propose the hypothesis of continental drift.

2. List the main objections to Wegener's hypothesis.

3. Define lithosphere and asthenosphere, then explain how they are related to the theory of plate tectonics.

4. Relate the concept of paleomagnetism to the Curie point and to polar wandering.

5. Describe Harry Hess's important hypothesis.

6. Explain the idea of geomagnetic reversals and discuss their relationship to the seafloor spreading hypothesis.

7. List the three basic types of plate boundaries and describe the relative motions at each.

8. Relate divergent boundaries to oceanic ridges, rift valleys, and seafloor spreading.

9. Name and describe what happens at each of the three types of convergent plate boundaries.

10. Describe the role of transform faults in the plate tectonics model.

11. Discuss the relationship between deep-focus earthquakes and the plate tectonics model.

12. List evidence gathered from the Deep Sea Drilling Project that helped validate the plate tectonics theory.

13. List and describe the three current models for explaining mantle convection and note why each model is imperfect.

14. Define and understand the key terms listed at the end of the chapter.

VOCABULARY REVIEW

1. About 200 million years ago, all of the continents on Earth were part of a single supercontinent called (a)_____. The hypothesis that explains how parts of this original supercontinent moved apart is called the hypothesis of (b)_____.

2. When iron-rich minerals cool below the (a)_____, they become magnetized in the direction parallel to the existing magnetic field. Rocks that possess such a magnetism from a time in the geologic past are said to possess a fossil magnetism, or (b)_____.

3. The hypothesis in which the seafloor is thought to move in conveyor-belt fashion is Hess's hypothesis of _____.

4. When rocks exhibit the same magnetic polarity as the present magnetic field, they are said to possess (a)_____ , but when rocks exhibit a polarity that is the opposite of the present magnetic field, they are said to possess (b)_____.

5. Sensitive instruments used to detect magnetic field polarity are called _____.

6. Rigid sections of lithosphere are called _____.

7. (a)_____ are narrow zones developed where plates are colliding (moving together). The place where two plates collide and an oceanic plate descends into the asthenosphere is known as a(n) (b)_____. As the oceanic plate descends, sediments and pieces of oceanic crust form a chaotic mass at the continental margin, which is called a(n) (c)_____.

8. _____ are narrow zones developed where plates are separating (moving apart).

9. _____ are narrow zones developed where plates are sliding past one another.

10. Large downfaulted valleys generated where plates are diverging are called _____ or _____.

11. Isolated points at which magma is upwelling from the mantle are called _____.

12. The region where an oceanic plate descends into the asthenosphere because of plate convergence is called a(n) (a)_____. This process begins when the oceanic plate bends downward to produce a deep linear trough on the seafloor, called a(n) (b)_____.

13. Mountains produced by volcanic activity associated with the subduction and melting of seafloor rocks and sediments are called (a)_____.
A chain of small volcanic islands formed in such a manner is called a(n) (b)_____.

14. _____ are prominent linear breaks in the oceanic crust that include transform faults and their inactive extensions into the plate interior.

15. According to the plate tectonics model, Earth's uppermost mantle and underlying crust behave as a rigid layer called the (a)_____. It overlies a weaker region of the mantle known as the (b)_____.

16. The (a)_____ hypothesis proposes that when cold, dense oceanic material is subducted, it pulls trailing lithosphere with it.

(b)_____ may occur when gravity sets in motion the elevated slabs astride ridge crests pushes the lithosphere in front of it.

APPLYING WHAT YOU HAVE LEARNED

1. On each blank provided below, write the name of the type of plate boundary illustrated.

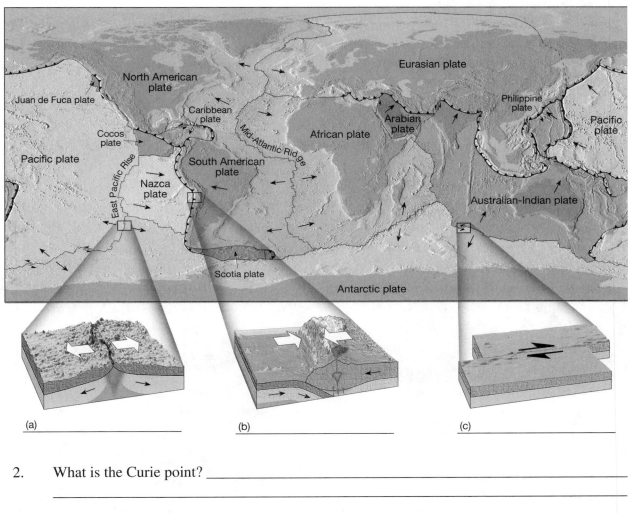

(a)_____   (b)_____   (c)_____

2. What is the Curie point? _____

_____

_____

_____

_____

3. What is reversed polarity? _____
_____
_____
_____
_____

4. What is a rift? _____
_____
_____
_____

5. On the blanks provided below, write the names of the features that are labeled.

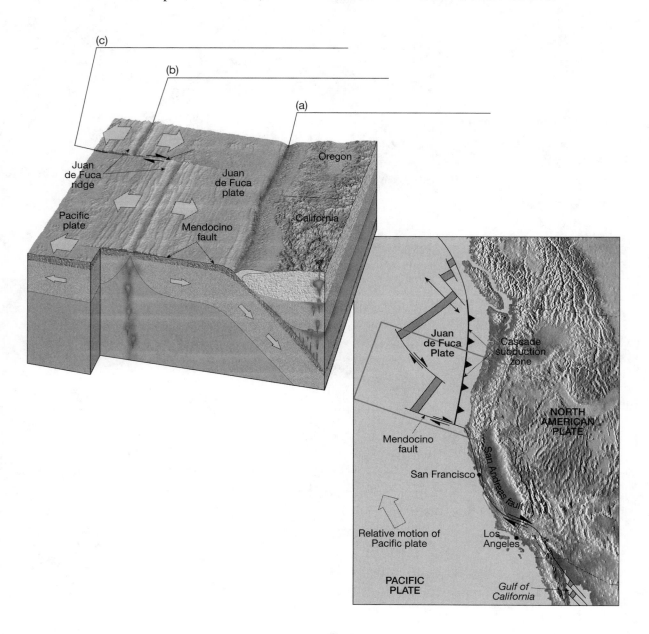

(c) _____

(b) _____

(a) _____

ACTIVITIES AND PROBLEMS

1. Define lithosphere and asthenosphere, then explain how they are related to the theory of plate tectonics. _____
   _____
   _____
   _____
   _____
   _____
   _____
   _____
   _____

2. How is the seafloor like a giant tape recorder? _____
   _____
   _____
   _____
   _____
   _____
   _____

3. List and describe what happens at each of the three types of convergent plate boundaries.
   _____
   _____
   _____
   _____
   _____
   _____
   _____
   _____
   _____

4. What evidence led Wegener to propose the hypothesis of continental drift? _____
   _____
   _____
   _____
   _____
   _____

5. Briefly describe Harry Hess's important hypothesis. _____

_____

_____

_____

_____

6. Explain the idea of geomagnetic reversals, and briefly discuss their relationship to the seafloor spreading hypothesis. _____

_____

_____

_____

_____

_____

_____

7. Name and briefly describe three hypotheses that were used before acceptance of the hypotheses of seafloor spreading and plate tectonics to explain the occurrence of similar species on land masses that are at present separated by vast oceans. _____

_____

_____

_____

_____

_____

_____

8. Name and describe the three currently proposed models for mantle convection, then describe why each model is imperfect. _____

_____

_____

_____

_____

_____

_____

_____

_____

REVIEW EXAM

1. What are the three main types of plate boundaries and the relative motions of plates associated with each type?

    (a)_____

    _____

    (b)_____

    _____

    (c)_____

    _____

2. What is the role of transform faults in the plate tectonics model? _____

    _____

    _____

    _____

    _____

3. Describe exactly what type of plate boundary is illustrated below and explain your answer.

    _____

    _____

    _____

    _____

    _____

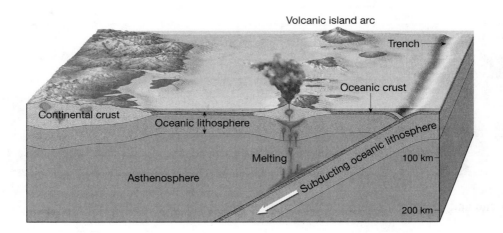

4. Recall the system of rift valleys and associated features that are presently developing in eastern Africa (Kenya). Briefly describe two main plate-tectonic stages of ocean-basin development that have already occurred (and are occurring) there.

First stage: (a)_____
_____
_____

Second stage: (b) _____
_____
_____

Predict what will happen to the large freshwater lakes of the present rift valleys in the geological future, and explain why this will happen. (c)_____
_____
_____
_____
_____
_____

5. How are deep-focus earthquakes related to the plate tectonics model? _____
_____
_____
_____
_____
_____

6. What evidence gathered from the Deep Sea Drilling Project has helped to validate the plate tectonics theory? _____
_____
_____
_____
_____
_____

7. Name and describe three hypotheses that may explain why plates move. _____
_____
_____
_____
_____
_____
_____
_____
_____

8.	Circle the letter of the most correct answer. Pangaea

	a.	means "all land."
	b.	was a supercontinent that existed about 200 million years ago.
	c.	did not include North America.
	d.	was both a and b.
	e.	was all of the above.

9.	Circle the letter of the most correct answer. When the continental edges of two lithospheric plates converge, the result is landforms of what type?

	a.	the Andes Mountains
	b.	volcanic island arcs
	c.	continental island arcs
	d.	the Himalayan Mountains

10.	Circle the letter of the most correct answer. Which item below is not found at divergent plate boundaries?

	a.	volcanoes
	b.	earthquakes
	c.	subduction zone
	d.	rift valleys

11.	Circle the letter of the most correct answer. Earth's present magnetic field is said to possess

	a.	normal polarity.
	b.	reversed polarity.
	c.	no polarity.
	d.	geomagnetic reversals.

12.	For each of the following statements, write T for *true* or F for *false* on the blank provided to indicate its validity.

	(a) _____	There are ocean trenches developed along the east coast of the United States.
	(b) _____	The Andes Mountains are an example of a continental volcanic arc.
	(c) _____	Transform faults occur only within ocean basins.
	(d) _____	Deep-focus earthquakes commonly occur below ocean ridge crests.

CHAPTER 20: MOUNTAIN BUILDING AND THE EVOLUTION OF CONTINENTS

LEARNING OBJECTIVES

After reading and studying this chapter, you should be able to

1.   Describe the basic structures characteristic of mountain chains developed throughout geologic history.

2.   Contrast development of Aleutian-type mountain building with Andean-type mountain building.

3.   Explain how mountains are produced by continental collisions.

4.   Describe the plate tectonic setting and process that produces fault-block mountains.

5.   Explain how orogenesis occurs with accretion.

6.   Explain the process of isostasy and the concept of isostatic adjustment and provide an example to illustrate the concept.

7.   Briefly describe the orogenic history of the Appalachian region for the past 600 million years.

8.   List the three main factors that affect the height of mountain chains.

9.   Describe George Airy's hypothesis.

10.   Contrast the two main opposing views on the origin and evolution of continents.

11.   Describe how convective flow in the mantle may contribute to development of domes and basins within plates.

12.   Define and understand the key terms listed at the end of the chapter.

VOCABULARY REVIEW

1.   The name for the processes that collectively produce a mountain system is

_____.

2.   Typically, mountain chains consist of numerous _____ that show evidence of being formed during the same mountain-building episode.

3.   The concept of crust floating on the mantle in gravitational balance is called (a)_____. If the thickness of crust changes and a new level of equilibrium is achieved, then (b)_____ has occurred.

4.   Sedimentary deposits scraped off a subducting plate pile up in front of the overriding plate to form what is known as a(n) _____.

5. A continental margin that is developed within a plate (rather than at a plate boundary) is called a(n) _____.

6. _____ are any crustal fragments whose geologic history is distinct from the adjoining crustal material.

7. Mountains caused by high- to moderate-angle normal faulting associated with tensional stresses are called _____.

APPLYING WHAT YOU HAVE LEARNED

1. What is isostasy? _____

_____

_____

_____

_____

2. What is isostatic adjustment of Earth's crust, and what causes it to occur? _____

_____

_____

_____

_____

_____

_____

_____

_____

_____

3. What is an accretionary wedge, and how does it form? _____

_____

_____

_____

_____

4. What is orogenesis? _____

_____

_____

_____

_____

ACTIVITIES AND PROBLEMS

1. The Piedmont section of the eastern Appalachian Mountains has a very subdued topography. Why is it included as part of the Appalachian Mountains? _____
_____
_____
_____

2. Some plateaus composed of horizontal strata have been deeply incised by streams to produce very rugged mountain-like landscapes. Are such landscapes mountains?
   (a)_____
   Explain. (b)_____
   _____
   _____
   _____
   _____

3. List the three main factors that affect the height of mountain chains (excluding erosion).
   _____
   _____
   _____

4. How do volcanic island arcs (Aleutian-type mountains) form? _____
   _____
   _____
   _____
   _____

5. How did the Himalayan Mountains form? _____
   _____
   _____
   _____
   _____
   _____

6. How did the Alps form? _____
   _____
   _____
   _____
   _____
   _____

REVIEW EXAM

1. What was George Airy's hypothesis? _____
    _____
    _____
    _____
    _____

2. Briefly describe how continental volcanic arcs (Andean-type mountains) form.
    _____
    _____
    _____
    _____
    _____

3. How do terranes probably form? _____
    _____
    _____
    _____
    _____

4. How many distinctly different episodes of mountain building have occurred in the
    Appalachian region over the past 600 million years? _____

5. Describe how convective flow in the mantle may contribute to development of domes and
    basins within plates. _____
    _____
    _____
    _____
    _____

6. Briefly stated, what are the two opposing views on the origin of continental crust?
    _____
    _____
    _____
    _____
    _____
    _____

7. Circle the letter of the most correct answer. After the top of a mountain belt is eroded, it will

    a.   rise higher to achieve isostatic adjustment.
    b.   sink to achieve isostatic adjustment
    c.   stay the same, because no orogenesis has occurred.

8.   For each of the following statements, write T for *true* or F for *false* on the space provided to indicate its validity.

(a) _____   The American Cordillera runs continuously from the tip of South America through Alaska.

(b) _____   The Himalayas are still rising orogenically.

(c) _____   Volcanic activity is commonly associated with the orogenesis of fault-block mountains.

(d) _____   Mountain belts are found on every continent.

(e) _____   A period of mountain building generally lasts about a million years.

CHAPTER 21: ENERGY AND MINERAL RESOURCES

LEARNING OBJECTIVES

After reading and studying this chapter, you should be able to

1. Distinguish between renewable and nonrenewable resources.

2. Describe the four main types of oil and gas traps.

3. Describe how burning coal can cause acid precipitation.

4. Predict two main potential sources of liquid fuels.

5. Describe how liquid fuels are produced from oil traps, tar sand deposits, and oil shales.

6. Outline the steps required to form anthracite coal.

7. List and describe six possible alternate energy sources.

8. Describe how nuclear energy is converted to electricity.

9. Name two reasons why the development of nuclear power has been hindered in recent years.

10. Describe how geothermal energy is produced.

11. Describe how hydroelectric power plants work to produce electricity.

12. Describe how tidal power is used to produce electricity.

13. List the three most favorable factors for a geothermal reservoir of commercial value.

14. Contrast vein deposits, disseminated deposits, and pegmatites.

15. Describe how enrichment by weathering processes creates valuable ore deposits.

16. Briefly describe how placers form.

17. Define the two broad groups of nonmetallic resources.

18. Briefly discuss the per capita consumption of metallic and nonmetallic resources.

19. List the main environmental problems that result from burning of fossil fuels.

20. Define and understand the key terms listed at the end of the chapter.

VOCABULARY REVIEW

1. Resources are classified as (a)_____ if they can be replenished over relatively short time spans. Resources are classified as (b)_____ if they are replenished so slowly that significant deposits of them take millions of years to accumulate.

2. Coal, oil, and natural gas are fuels formed from the remains of ancient plants and animals, so they are collectively called _____.

3.	(a)_____ are all of the minerals that will ultimately be available commercially for use by people. More specifically, minerals containing metals are regarded as (b)_____, whereas nonfuel and nonmetallic types of minerals that will ultimately be used by people are regarded as (c)_____.

4.	A geologic environment that allows for economically significant amounts of oil and gas o accumulate is called a(n) (a)_____, which always consists of a permeable (b)_____ overlain by (c)_____ _____ such as shale.

5.	Nuclear power plants use radioactive fuels that release energy by the process of _____.

6.	_____ are mineral resources from which minerals can be profitably extracted.

7.	_____ is the name applied to useful metallic minerals that can be mined at a profit.

8.	_____ are igneous rocks composed of unusually large crystals.

9.	Hot, metal-rich fluids that are remnants of late-stage magmatic processes are called _____.

10.	Accumulations of mineral resources found in fractures that have been filled with minerals are called (a)_____. Accumulations of ore minerals that are finely distributed as minute masses throughout the entire rock mass are called (b)_____ _____.

11.	_____ is said to occur when weathering concentrates minor amounts of metals that were scattered throughout unweathered rock, so that ores are produced.

12.	_____ form when heavy-mineral grains are mechanically sorted and concentrated by currents.

13.	Electric power generated by using falling water to drive turbines is called_____ _____ power.

APPLYING WHAT YOU HAVE LEARNED

1. What is a nonrenewable resource? _____
   _____
   _____
   _____

2. What is an oil trap? _____
   _____
   _____
   _____

3. What is an ore? _____
   _____
   _____
   _____

4. What is a secondary enrichment? _____
   _____
   _____
   _____
   _____

5. What is a placer? _____
   _____
   _____
   _____
   _____

ACTIVITIES AND PROBLEMS

1. List the three primary fuels of our modern industrial economy. _____
   _____
   _____
   _____

2. Briefly outline the steps that lead to the formation of anthracite from plants. _____
   _____
   _____
   _____
   _____
   _____
   _____

3. On the blank provided beside each illustration below, write the name of the kind of oil and gas trap that is shown.

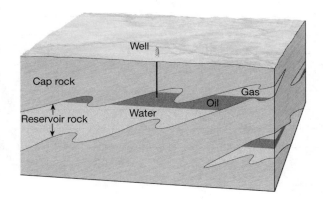

(a) _____

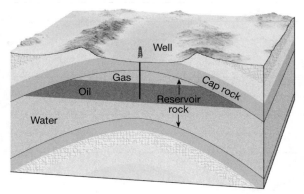

(b) _____

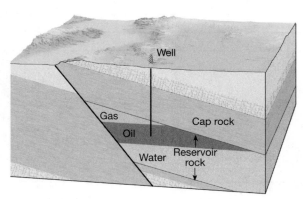

(c) _____

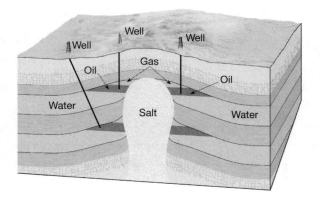

(d) _____

4. What is the major difference in the way that liquid fuels are obtained from conventional oil reservoirs versus tar sand deposits? _____

_____

_____

_____

_____

5. Describe how burning of coal causes acid precipitation to occur. _____

_____

_____

_____

_____

6. What are three ways that solar energy is used to heat buildings? _____

_____

_____

_____

_____

7. Describe how geothermal energy is produced. _____

_____

_____

_____

_____

_____

_____

REVIEW EXAM

1. How is oil extracted from oil shales? _____

_____

_____

_____

_____

_____

2. Briefly describe the basic steps involved in converting nuclear energy to electricity.

   _____
   _____
   _____
   _____
   _____
   _____
   _____

3. How is hydroelectric power produced? _____

   _____
   _____
   _____
   _____
   _____

4. How is tidal power used to produce electricity? _____

   _____
   _____
   _____
   _____
   _____

5. What are the three most favorable factors for a geothermal reservoir of commercial value?

   _____
   _____

6. What are the two main groups of nonmetallic resources? _____

   _____

7. List six possible alternate energy sources. _____

   _____
   _____
   _____
   _____
   _____
   _____

8.   Circle the letter of the most correct answer. Two of the most likely future sources of liquid fuels are

   a.   limestone caves and tar pits.
   b.   tar sands and the moon.
   c.   oil shales and tar sands.
   d.   oil shales and the moon.
   e.   the moon and Mars.

9.   Circle the letter of the most correct answer. Roughly half of the world's supply of oil shale is located in

   a.   Colorado, Utah, and Wyoming.
   b.   California and Nevada.
   c.   Alaska.
   d.   Alberta, Canada.

10.   Circle the letter of the most correct answer. Bauxite is produced by

   a.   an igneous process.
   b.   a metamorphic process.
   c.   secondary enrichment.
   c.   sedimentary sorting.

11.   Circle the letter of the most correct answer. Copper and gold are

   a.   nonmetallic resources.
   b.   renewable resources.
   c.   nonrenewable resources.
   d.   industrial minerals.
   e.   both c and d.

12.   Circle the letter of the most correct answer. The percentage of U.S. energy demand met collectively by solar power, geothermal power, wind energy, and tidal power is about

   a.   90%
   b.   41%
   c.   4%
   d.   1%

13.   Circle the letter of the most correct answer. The main fuel used to generate electricity in the U.S. is

    a.   coal.
    b.   nuclear.
    c.   petroleum.
    d.   natural gas.

14.   For each of the following statements, write T for *true* or F for *false* on the blank provided to indicate its validity.

    (a) _____   Underground mining is now highly mechanized, so it is no longer dangerous.
    (b) _____   The hydrocarbon material in tar sands is kerogen.
    (c) _____   The primary fuel used in nuclear power plants is plutonium-235.
    (d) _____   Nuclear power plants cannot explode like an atom bomb.
    (e) _____   The per capita consumption of metallic mineral resources is greater than the per capita consumption of nonmetallic mineral resources.
    (f) _____   Some nonmetallic mineral resources include gypsum, halite, and sulfur.
    (g) _____   Some metallic mineral resources include gold, zinc, copper, and platinum.

CHAPTER 22: PLANETARY GEOLOGY

LEARNING OBJECTIVES

After reading and studying this chapter, you should be able to

1. Describe the differences between Jovian planets and terrestrial planets of our solar system, and list the members of each group.

2. List and distinguish among the three types of materials that make up the planets.

3. Briefly describe the events which are thought to have led to the formation of our solar system.

4. Name the two types of lunar terrain first identified by Galileo.

5. Outline the steps in the evolution of the moon.

6. Describe Mercury in terms of size, density, surface temperatures, and surface terrain.

7. Explain why Venus is referred to as "Earth's twin."

8. Explain the primary reason for the blistering surface temperatures on Venus.

9. List some Martian surface features that are also common on Earth.

10. List ideas on the origin of Martian valleys that do not have a tectonic origin.

11. Briefly describe the structure of Jupiter's atmosphere.

12. Compare the four Galilean satellites of Jupiter and briefly contrast them with Jupiter's four outermost satellites.

13. Describe the evidence indicating that Saturn's rings are composed of individual moonlets rather than solid disks.

14. Explain why Uranus and Neptune are often called "twins."

15. Explain the differences among meteoroids, meteors, and meteorites.

16. Briefly describe the structure and relative size of Pluto.

17. Infer what would happen on Earth if it passed through the tail of a comet.

18. Define and understand the key terms at the end of the chapter.

VOCABULARY REVIEW

1. The large Jupiter-like planets of our solar system that have relatively low densities and thick atmospheres are called (a)_____ planets. The smaller Earth-like planets of our solar system that have relatively high densities and thin atmospheres are called (b)_____ planets.

2. The _____ suggests that all bodies of our solar system formed from an enormous cloud consisting of about 80% hydrogen, 15% helium, and 5% of the other elements.

3. The two main features of the moon first recognized by Galileo were the dark, flat regions called (a)_____ and the heavily cratered (b)_____.

4. Lunar rock fragments and dust (from impact cratering) that have been welded together form a rock called _____.

5. The linear "splash-marks" that radiate outward from some craters are called _____.

6. "Lunar soil" should actually be called _____.

7. The large space in Saturn's rings (between the A-ring and B-ring) is called the _____.

8. _____ is the disappearance of light resulting when one object passes behind an apparently larger one.

9. Small planet-like bodies that are only 1–100 km across are called _____ _____.

10. A(n) (a)_____ is a small body that revolves around the sun in an elongated orbit and develops a tail as it approaches the sun. The glowing head of such a body is called the (b)_____.

11. Any small solid particle in orbit within our solar system is called a(n) (a)_____ _____. Swarms of such particles occasionally enter the Earth's atmosphere and create spectacular glowing displays called (b)_____. If remains of these glowing particles are found on the Earth's surface, then they are called (c)_____ _____. Occasionally, these remains are dust-sized particles called (d)_____.

12. Meteorites composed chiefly of silicate minerals are classified as (a)_____ _____. Those composed chiefly of iron and nickel are classified as (b)_____ _____, and those composed chiefly of a mixture of silicate minerals and iron are classified as (c)_____.

APPLYING WHAT YOU HAVE LEARNED

1.   What is a Jovian planet? _____
      _____
      _____

2.   What is a terrestrial planet? _____
      _____
      _____
      _____

3.   What is a stony-iron meteorite? _____
      _____
      _____
      _____

4.   What is a meteoroid? _____
      _____
      _____
      _____

5.   What is a meteor? _____
      _____
      _____
      _____

6.   What are asteroids? _____
      _____
      _____
      _____

7.   What is lunar regolith? _____
      _____
      _____

8.   What is a coma? _____

ACTIVITIES AND PROBLEMS

1.   List and distinguish among the three types of materials that make up planets.
      _____
      _____
      _____

2. List the Jovian planets of our solar system, from nearest to farthest from the sun.

    _____

    _____

    _____

3. Carefully examine the telescopic view of the lunar surface (near side) below and the features labeled A, B, and C.

Give a brief description that characterizes the bright area around C. (a)_____

_____

_____

Notice the features like those labeled A. What are such features called?
(b)_____
How are craters like the one labeled B different from the craters of the area labeled C?
(c)_____

_____

_____

_____

In what order (from first to latest) did the features form that are labeled A, B, and C?
(d)_____
Briefly describe the step-by-step sequence of events that led to the development of the
lunar terrain as it appears today (being sure to include events that led to development of
features like A, B, and C. (e)_____

_____

_____

_____

_____

_____

_____

_____

_____

4.   Briefly describe the events that led to the formation of our solar system. _____

_____

_____

_____

_____

_____

_____

5.   What is the main reason why Venus has blistering surface temperatures? _____

_____

_____

_____

6.   Why does the Earth have only a small amount of carbon dioxide, but much free oxygen, in its
atmosphere? _____

_____

_____

_____

7. Carefully examine the photograph below of the Martian landscape. Notice the dark-colored rocks. Also note the deposits of light-colored sand, especially at the left-hand side of the photograph.

(Photo courtesy of NASA)

On the basis of this picture, describe the surface and atmospheric conditions at this location on Mars? (a)_____

_____

_____

_____

_____

How is this typical picture of the Martian landscape different from a typical picture of an Earth landscape? (b)_____

_____

_____

_____

_____

8. What would happen on Earth if it passed through the tail of a comet? _____

_____

_____

_____

_____

REVIEW EXAM

1. Why is Venus referred to as "Earth's twin"? _____

_____

_____

_____

2. Describe the size, density, surface temperatures, and surface terrain of Mercury.

_____
_____
_____
_____
_____
_____
_____

3. Briefly describe what causes the alternating light- and dark-colored bands of Jupiter's atmosphere. _____

_____
_____
_____
_____

4. Briefly describe the structure and composition of Pluto. _____

_____
_____
_____
_____
_____

5. What are the two types of lunar terrain that were first identified by Galileo? _____

_____
_____
_____

6. List several surface features of Mars that are also common on Earth. _____

_____
_____
_____
_____
_____
_____

7. What are at least two different ideas on the origin of Martian valleys that do not have a tectonic origin? _____

_____
_____
_____
_____
_____
_____

8. Why are Uranus and Neptune often called "twins"? _____
_____
_____
_____
_____
_____
_____

9. Circle the letter of the most correct answer. How does the density of Mars's atmosphere compare with the density of Earth's atmosphere?

   a. Mars's atmosphere is less dense.
   b. Both atmospheres have about the same density.
   c. Mars's atmosphere is more dense.

10. Circle the letter of the most correct answer. The Jovian planet below is

    a.   Earth.
    b.   Venus.
    c.   Mercury.
    d.   Saturn.

11. Circle the letter of the most correct answer. The brightest planet observed in Earth's night sky is

    a.   Venus.
    b.   Mars.
    c.   Mercury.
    d.   Saturn.

12. For each statement below, write T for *true* or F for *false* on the space provided to indicate its validity.

    (a) _____   Pluto is a Jovian planet.
    (b) _____   Saturn's density is less than the density of water.
    (c) _____   A planet's ability to retain an atmosphere is entirely dependent on its surface temperature.
    (d) _____   The "back side" of the moon has highlands and maria, and so does the "front side."
    (e) _____   The largest planet in our solar system is Neptune.

200

ANSWER KEY

CHAPTER 1: AN INTRODUCTION TO GEOLOGY

1. a. Earth.

VOCABULARY REVIEW (Chapter 1)
1. a. hydrosphere; b. atmosphere
2. a. inner core; b. outer core; c. mantle; d. crust
3. the study of Earth, or literally, from the Greek, "Earth discourse"
4. a. Physical; b. Historical
5. catastrophism
6. a. the physical, chemical, and biological laws that operate today have also operated in the geologic past; b. James Hutton
7. system
8. a. 4004; b. 4.5 billion (4,500 million)
9. a. Relative; b. in an undeformed sequence of sedimentary rocks or lava flows, each layer is older than the one above it and younger than the one below it; c. fossil succession.
10. vast cloud of dust and gases in space
11. a. hypothesis or model; b. theory
12. nebular hypothesis
13. a. asthenosphere; b. mesosphere or lower mantle; c. lithosphere
14. a. oceanic ridge system; b. trenches
15. a. plate tectonics; b. plates; c., d., and e. (in any order): divergent, convergent, transform fault
16. a. seafloor; b. spreading
17. shields
18. subduction zones
19. a. magma; b. igneous
20. a. weathering; b. sediment; c. lithification; d. sedimentary
21. metamorphic
22. a. continental shelf; b. continental slope
23. paradigms
24. a. hydrosphere; b. atmosphere; c. biosphere; d. solid Earth

APPLYING WHAT YOU HAVE LEARNED (Chapter 1)
1. It means that our planet consists of many interacting parts that form a complex whole.
2. Collect facts; develop a scientific hypothesis; construct experiments to test the hypothesis; accept, modify, or reject the hypothesis based on extensive testing.
3. a. continental shelf; b. continental slope; c. trench.
4. a. convergent; b. divergent; c. convergent; d. subduction zone; e. transform faults.
5. a. A huge rotating cloud of dust and gases (nebula) began to contract.
   b. Most of the dust and gases were swept toward the center, producing the sun. However, some dust and gas remained orbiting the sun in a flattened disk.
   c. Planets accreted from material within the flattened disk.
   d. In time, most of the remaining debris was either accreted into the nine planets and their moons or it was swept into space by the solar wind.
6. a. crystallization (cooling and solidification); b. weathering (decomposition and disintegration); c. lithification (compaction and cementation); d. metamorphism (reaction to heat and pressure); e. melting

ACTIVITIES AND PROBLEMS (Chapter 1)

1. The law of superposition and the pattern of fossil succession were used to subdivide and name the layers, thereby making a relative time scale. Much later, ages in years were determined for some layers and added to complete the time chart.
2. Both doctrines have been extensively tested and accepted as laws. Also, it is because gradual, subtle processes that go unnoticed in a lifetime can have an effect over geologic time that equals the effect of one great, sudden catastrophe.
3. It is because geologic spans of time far exceed the span of a human lifetime.
4. Fossils of humans are found only in geologically younger rocks.
5. Most of the air in Earth's atmosphere probably came from the planet's interior (a process called degassing), although the free oxygen probably came from plants. Almost none of the air is gas remaining from the primordial nebula, because that gas was swept away by the solar wind and incorporated into the outer planets.
6. a. lakes, streams, oceans, glaciers, etc.; b. They are all part of the water (hydrologic) cycle.
7. The trench marks the location where seafloor lithosphere descends into a subduction zone. As the subducting lithosphere starts to melt in the subduction zone and also steam the base of continental lithosphere, magma forms and rises to develop the volcano.
8. No. The rock cycle is oversimplified to show the ideal cycle, even though there are many alternative paths in the real world. For example, any rocks exposed at the Earth's surface will weather and produce sediment.
9. The natural world behaves in a consistent and predictable manner.
10. The sun drives external processes. Heat from Earth's interior drives internal processes.

REVIEW EXAM (Chapter 1)

1. They have been destroyed by plate tectonic processes, weathering, and erosion.
2. trenches
3. d
4. d
5. a
6. b
7. a
8. b
9. a
10. a. False. The densest portion of the earth is its inner core; b. True; c. False. Shields are indeed the oldest features of the earth's surface, but seafloors are geologically much younger; d. True; e. True; f. False. The positions of the shorelines have varied much over geologic time; g. True; h. False. When continents collide, majestic folded mountains are formed; i. False. Erosion is a destructional process, but volcanism is a constructional process; j. True
11. a. lithosphere; b. asthenosphere; c. crust; d. mantle; e. outer core; f. inner core

CHAPTER 2: MATTER AND MINERALS

VOCABULARY REVIEW (Chapter 2)

1. mineralogy
2. a. elements; b. rocks
3. polymorphs
4. crystal form
5. atom
6. luster
7. streak
8. a. nucleus; b. protons; c. neutrons; d. electrons

9. a. atomic number; b. atomic mass
10. isotopes
11. oxygen, silicon, aluminum, iron, calcium, sodium, potassium, magnesium
12. a. ion; b. ionic
13. a. hardness; b. Mohs
14. a. cleavage; b. fracture
15. compound
16. specific gravity
17. radioactivity
18. a. energy-level shells; b. 8
19. valence
20. a. covalent; b. metallic
21. ferromagnesian
22. rock
23. a. silicates; b. silicon-oxygen tetrahedron
24. a. 3-dimensional network; b. sheet; c. 3-dimensional network; d. single tetrahedron; e. chains; f. double chains
25. a. gypsum; b. halite; c. graphite (Some pencil leads also contain clay from clay minerals, but the main mineral is graphite.); d. bauxite; e. talc

APPLYING WHAT YOU HAVE LEARNED (Chapter 2)
1. a. No, because it has equal numbers of protons and electrons; b. 2; c. 2; d. 6; e. 8; f. 2
2. a. Yes, because it has 17 protons and 18 electrons, for a net charge of -1; b. 0; c. 3
3. a. 3; b. 3; c. 4
4. A mineral is a substance that has a definite chemical composition, has atoms arranged in a definite pattern (has orderly internal structure), is an inorganic solid, and occurs naturally.
5. It has a conchoidal fracture.
6. Cleavage is the tendency for minerals to break along flat surfaces that are planes of weak bonding in their orderly internal structure.

ACTIVITIES AND PROBLEMS (Chapter 2)
1. Yes. It is a naturally occurring, inorganic solid, with a specific chemical formula and definite internal structure.
2. No. Sugar is organic (made by plants), and minerals are inorganic.
3. The ferromagnesian minerals contain iron (Fe) and/or magnesium (Mg) and are dark in color. The nonferromagnesian minerals lack Fe and Mg and are light in color.
4. a. 14; b. 12; c. yes; d. because they both have 6 protons, and the number of protons identifies the atom.
5. 8.9
6. a. Na (sodium); b. because Na and Ca atoms are about the same size; c. the ratio of Si, Al, and O relative to the Na
7. Blue and brown asbestos fibers are thin, rigid, rod-like needles that easily penetrate tissues, but white Chrysotile asbestos fibers are soft, curly, and do not penetrate tissues.

REVIEW EXAM (Chapter 2)
1. Mg (magnesium)
2. one
3. plagioclase feldspar
4. olivine
5. amphibole
6. rarity
7. quartz (SiO_2)

8. Karat is a term used to indicate purity of gold; whereby pure gold is 24 karat (24K) gold. A carat is a unit of weight (0.2 grams) for gemstones.
9. radioactivity
10. a. nonferromagnesian (light); b. It is light in color and its specific gravity is lower than ferromagnesians.
11. Polymorphs are minerals that have the exact same chemical composition but different physical properties.
12. feldspar
13. clay minerals
14. quartz
15. b
16. a
17. b
18. c
19. a. True; b. False. Calcite has a hardness of 3, but talc has a hardness of only 1; c. False. They have different numbers of neutrons but the same number of protons; d. False. It would have three energy-level shells; e. False. Streak is the color of a mineral in powdered form. Luster is the way a surface of a mineral reflects light; f. False. For example, quartz can have excellent crystal form, but it has no visible cleavage; g. True
20. a. none; b. slightly more than 3:1 (22:7 or 3.14 :1); c. pyroxene group (augite); d. slightly less than 3:1 (41:14 or 2.92:1); e. amphibole group (hornblende); f. two planes at 60 and 120 degrees

CHAPTER 3: IGNEOUS ROCKS

VOCABULARY REVIEW (Chapter 3)
1. gases and suspended crystals
2. a. and b. (either order): intrusive; plutonic
3. a. and b. (either order): extrusive; volcanic
4. crystallization
5. a. glassy; b. glass or obsidian
6. a. phaneritic; b. aphanitic
7. crystal settling
8. a. vesicles; b. vesicular
9. decompressional melting
10. a. porphyry; b. porphyritic; c. phenocrysts; d. groundmass
11. pyroclastic or fragmental
12. Bowen's reaction series
13. a. granite; b. rhyolite
14. basalts
15. pumice
16. magmatic differentiation
17. pegmatitic
18. magma mixing
19. geothermal gradient
20. partial melting

APPLYING WHAT YOU HAVE LEARNED (Chapter 3)
1. a. peridotite (phaneritic ultramafic rock); b. gabbro (phaneritic mafic rock); c. basalt (aphanitic mafic rock); d. diorite (phaneritic intermediate rock); e. andesite (aphanitic intermediate rock); f. granite (phaneritic felsic rock); g. rhyolite (aphantic felsic rock)

ACTIVITIES AND PROBLEMS (Chapter 3)

1. Magma is molten rock deep beneath the land surface, whereas lava is molten rock at or near the land surface.
2. In general, fast cooling of magma produces small crystals (aphanitic, fine-grained texture). Slow cooling provides time for larger crystals to form (phaneritic, coarse-grained texture).
3. In magma with low viscosity (low resistance to flow; flows easily), ions can move about freely and arrange themselves into an orderly crystalline pattern. In magma with high viscosity (high resistance to flow; flows poorly), ions cannot move about freely and cannot arrange themselves into an orderly crystalline pattern. A glass (volcanic glass, obsidian) forms instead of crystals.
4. a. Textural name: porphyritic. Origin: The magma cooled slowly at first (and the large phenocrysts formed), but then cooled quickly (to form the black groundmass of fine-grained crystals or glass.
   b. Textural name: aphanitic (fine-grained). Origin: The magma or lava cooled quickly, so only tiny crystals had time to form.
   c. Textural name: glass or glassy. Origin: The magma cooled suddenly and/or was too viscous for ions in the magma to arrange themselves into a crystalline pattern.
   d. Textural name: vesicular and aphanitic. Origin: The magma or lava cooled quickly, so only tiny crystals had time to form (aphanitic texture). Bubbles of gas (volatiles) were "frozen" in the cooling magma/lava to form the vesicles.
5. a. andesitic (intermediate); b. andesite
6. N. L. Bowen discovered that certain minerals crystallize first from a magma, that the composition of the magma continually changes as crystallization occurs, and that crystals remaining in magma after they form will react with that magma to produce the next mineral in the sequence.
7. Magma can be generated by increasing the rock's temperature until it starts to melt. Decreasing the confining pressure around the rock can cause decompressional melting. Introduction of volatiles (water) can lower the rock's melting point enough to start melting the rock.

REVIEW EXAM (Chapter 3)

1. a
2. c
3. a
4. d
5. d
6. a
7. a. True; b. True; c. True; d. False. Ca-rich plagioclase forms before Na-rich plagioclase until the available Ca is used up (at which point Na-rich plagioclase begins to form); e. True; f. False. The continuous reaction series is a plagioclase feldspar series. g. False. Granite is light colored, so it is a felsic rock type.

CHAPTER 4: VOLCANIC AND PLUTONIC ACTIVITY

VOCABULARY REVIEW (Chapter 4)

1. resistance to flow or mobility
2. a. pahoehoe; b. aa
3. Pillow
4. a. pyroclastic; b. lahar
5. a. crater; b. caldera
6. a. parasitic cones; b. fumeroles
7. a. shield volcanoes; b. cinder cones or scoria cones
8. volcanic domes
9. a. and b. (either order): composite cones; stratovolcanoes
10. nuée ardente

11. volcanic neck
12. pipes
13. a. fissure; b. flood basalts; c. pyroclastic flows
14. a. discordant; b. concordant
15. xenoliths
16. volatiles
17. lava tubes
18. fissure
19. a. volcanic island arc or volcanic arc; b. intraplate volcanism

APPLYING WHAT YOU HAVE LEARNED (Chapter 4)
1. Hot spots are localized volcanic regions where volcanism often persists for millions of years. They form where plumes of magma rise directly from deep within Earth's mantle.
2. The higher the percentage of silica (Si) in magma, the greater its viscosity. The lower the percentage of silica in magma, the lower the viscosity.
3. The greater the amount of dissolved gases in the magma, the less its viscosity. The lower the amount of dissolved gases, the greater the viscosity.
4. a. cinder; b. shield; c. composite
5. It is because many subduction zones and volcanoes border the Pacific basin.
6. a. composite cone; b. It formed by the accumulation of alternating layers of lava and pyroclastic material.
7. a. volcano, D, M; b. lava flow, C, T; c. sill, C, T; d. dike, D, T; e. laccolith, C, M; f. batholith, D, M
8. pahoehoe
9. lapilli
10. the magma's viscosity, the amount of dissolved gases it contains, and the ease with which gases can escape
11. a. Minor earthquakes (tremors) occurred directly beneath the mountain; b. movements of magma
12. The upper portion of a buried lava flow contains voids (vesicles) produced by entrapped gas bubbles, and only the rocks beneath the lava flow show evidence of metamorphic alteration (contact metamorphism). Sills form when magma has been forcefully intruded between layers of bedrock, so there is evidence of metamorphic alteration (contact metamorphism) of the rocks both above and below the sill.

ACTIVITIES AND PROBLEMS (Chapter 4)
1. Granitic magma is less dense than the rocks on top of it and around it, so its buoyancy causes it to migrate upward (like the "lava" warmed in a lava lamp). When the rising magma encounters cool, brittle rock, it fractures and dislodges blocks of the rock. This allows the magma to rise again until it cools and can no longer move.
2. size, shape (tabular or massive), and orientation relative to existing structures (discordant versus concordant)
3. Gases can easily and continuously escape from fluid basaltic lavas, but gases are trapped in very viscous magmas until they form large bubbles or pockets that may explode.

REVIEW EXAM (Chapter 4)
1. a. Columbia River basalts (or flood basalts); b. Extensive fissure eruptions led to development of flood basalts that covered parts of three states; c. The region was located over a hot spot when the flood basalts formed.
2. c
3. a. volcanic bomb; b. Lava ejected into the air forms a tear-drop shape and cools to form a rocky volcanic bomb, which falls to the ground.
4. a. True; b. False. It is about 20–30° C/km; c. False. They are andesitic; d. True; e. False. It is located over a subduction zone; f. False; g. True; h. False. They cut across such surfaces; i. True

5. a. xenolith; b. The dark-colored rock formed first. It was then incorporated into a body of magma that cooled to form the light-colored rock.
6. a. crater; b. pipe or conduit
7. a. A violent eruption emptied the magma chamber; b. The volcano collapsed because of the empty magma chamber beneath it; c. Rainfall, melting snow, and ground water drained into the depression to form the lake, but not until a very minor eruption produced Wizard Island.
8. d
9. a
10. b
11. b
12. c
13. a
14. c

CHAPTER 5: WEATHERING AND SOIL

VOCABULARY REVIEW (Chapter 5)
1. weathering
2. frost wedging
3. erosion
4. a. Chemical weathering; b. mechanical weathering
5. mass wasting
6. hydrolysis
7. spheroidal weathering
8. regolith
9. joints
10. Oxidation
11. dissolution
12. soil
13. talus slopes
14. a. sheeting; b. exfoliation domes
15. a. horizons; b. soil profile
16. eluviation
17. humus
18. a. internal; b. external
19. leaching
20. parent material
21. solum
22. a. pedalfers; b. pedocals
23. differential

APPLYING WHAT YOU HAVE LEARNED (Chapter 5)
1. It is the process by which blocks of rock weather to more rounded shapes, because the corners of blocks weather faster than the sides of the blocks.
2. Acid precipitation is precipitation that is more acidic than normal. It causes lakes and streams to be more acidic (which kills fish), promotes corrosion of metals, and promotes decomposition of stone structures and rocks.
3. a. joints; b. frost wedging and chemical weathering; c. spheroidal weathering
4. Caliche is a calcite (calcium carbonate) enriched layer in soil. It forms from the evaporation of soil water containing dissolved calcium carbonate. When the water evaporates, the calcium carbonate (calcite) precipitates as caliche.

5. C horizon
6. Oxidation is the process by which rusting occurs because an element loses electrons to react with (bond with) oxygen.
7. a. talus slope; b. Rocks fell from the adjacent rocky cliff and accumulated at the base of the cliff as a talus slope.

ACTIVITIES AND PROBLEMS (Chapter 5)
1. a. calcite sand weighing 10 grams; b. because there is more surface area available for chemical reaction
2. Minerals formed at higher temperatures decompose faster than those formed at lower temperatures. Therefore, silicate minerals decompose in the order in which they formed, according to Bowen's reaction series.
3. a. D; b. A; c. B; d. B
4. Hydrogen ions attack and replace positive ions in crystal lattices. This destroys the original orderly arrangement of atoms, so the minerals decompose.
5. The hotter and more humid the climate, the faster the rate of weathering. The cooler and drier the climate, the slower the rate of weathering.

REVIEW EXAM (Chapter 5)
1. frost wedging, unloading, thermal expansion, biological activity
2. The longer a soil has been forming, the thicker it becomes and the less it resembles its parent material.
3. Angular topography is more common in arid regions, whereas subdued topography is more common in humid areas. This is because the effects of chemical weathering and soil development are greater in humid areas than in arid areas.
4. Carbonic acid forms when carbon dioxide is dissolved in water (such as a droplet of rain).
5. a. iron, aluminum, clay minerals; b. calcium carbonate (calcite); c. tropical (heavy rainfall); d. brick red
6. type of parent material, climate, time, plants and animals, slope of the land
7. d
8. b
9. c
10. c
11. b
12. The soils present beneath such rainforests are laterites: soils containing essentially no soluble materials. Laterites are very poor soils for farming.
13. Strong wind, drought, and the transformation of grass-covered prairies into farms all contributed to the Dust Bowl.
14. a. False. It even dissolves in very weakly acidic solutions; b. True; c. True; d. False. It is called humus, or the O horizon; e. False. Soil profiles on steep slopes are typically developed very poorly.
15. a. O horizon; b. A horizon; c. B horizon; d. C horizon

CHAPTER 6: SEDIMENTARY ROCKS

VOCABULARY REVIEW (Chapter 6)
1. a. Sediment; b. Lithification
2. detrital
3. chemical
4. a. beds or strata; b. bedding planes
5. compaction
6. a. graded beds; b. cross-bedding

7. Ripple marks
8. Mud cracks
9. sorting
10. a. Fissility; b. shale
11. fossils
12. a. salt flats; b. evaporites
13. a. cementation; b. cement
14. clastic
15. nonclastic (or crystalline)
16. facies
17. sedimentary environment or environment of deposition
18. Diagenesis

APPLYING WHAT YOU HAVE LEARNED (Chapter 6)

1. Fossils are any remains or traces of prehistoric life.
2. bedding (strata): layers of sediment.
   bedding planes: surfaces between layers of sediment.
   cross-bedding: strata/beds deposited in an inclined position.
   graded bedding: sedimentary layers with particle sizes that fine upwards.
   ripple marks: small waves of sand.
3. A facies is a rock unit having distinctive characteristics that reflect the conditions of a particular environment in which it was formed.
4. It is a process by which unconsolidated (loose) sediment is transformed into solid sedimentary rock.
5. Dolomite (dolostone) contains Mg [it is $CaMg(CO_3)_2$], and limestone does not [it is $CaCO_3$].
6. Chert is a dense, hard rock made of microcrystalline silica (SiO_2).
7. a. mud cracks; b. Mud shrinks and cracks as it dries (as in an evaporating mud puddle).
8. a. Name: conglomerate (a clastic, detrital rock). Origin: *A poorly sorted mixture of rounded grains* of sand and gravel was lithified.
   b. Name: biochemical limestone or coquina. Origin: A mass of calcareous seashells and broken shells was cemented together.
   c. Name: breccia (a clastic, detrital rock). Origin: *A poorly sorted mixture of angular grains* of sand and gravel was lithified.
9. a. greater than 2; b. 1/16 to 2; c. less than 1/16
10. They are sedimentary rocks composed of transported solid particles (grains of sediment) that were derived from both mechanical and chemical weathering.
11. They are sedimentary rocks composed of materials that were physically or organically (biochemically) precipitated from solutions containing dissolved substances.
12. Fossil fuels (coal, oil, gas, tar)—Organisms used carbon dioxide from the atmosphere and the process of photosynthesis to make their food and soft tissues. These soft tissues were fossilized as fossil fuels. Limestone—Organisms used carbon dioxide from the atmosphere or hydrosphere to make their calcite and aragonite shells, which became sediment that was eventually cemented to form the limestone.
13. Carbonic acid in rainwater dissolves limestone and releases carbon to the hydrosphere. Also, when fossil fuels are burned or oxidize, they release carbon dioxide to the atmosphere. There, it mixes with water to make carbonic acid in rainwater, which falls to Earth's surface and enters the hydrosphere.

ACTIVITIES AND PROBLEMS (Chapter 6)

1. a. nonclastic; b. any size; c. chemical; d. clastic; e. 1/16 to 2 mm; f. detrital; g. clastic; h. greater than 2 mm; i. detrital; j. clastic or nonclastic; k. any size; l. chemical or biochemical; m. nonclastic; n. any size; o. chemical
2. Sorting refers to the degree of similarity in particle size in a sedimentary rock. It occurs as wind or water blow and wash the different sizes of particles apart.

3. If a great volume of plants accumulates in an oxygen-lacking environment (such as the bottom of a swamp or bog), then the plants do not decompose. As these plant deposits are buried by more plants, they turn into a brown spongy mass called peat. With additional burial and time, the peat undergoes chemical and physical changes that transform it into coal.
4. Cementation occurs when cementing materials are carried into solution by water percolating through open spaces in sediment. Through time, the cement precipitates onto the sedimentary particles, fills the open spaces, and cements the particles together. The most common cementing materials are calcite, iron oxide, and silica.

REVIEW EXAM (Chapter 6)
1. Compaction occurs in old sedimentary layers as new layers oil on top and squeeze the old layers to a more compact and hard form. Cementation occurs when mineral crystals crystallize in the pore spaces of bodies of sediment, where they "glue" together the grains of sediment.
2. As preexisting rocks weather at the Earth's surface, sedimentary particles are derived, eroded, transported, and deposited. The sedimentary deposits are then lithified to form detrital sedimentary rocks. Some earth materials are also dissolved and transported by water until they precipitate to form chemical sedimentary rocks or cement in detrital sedimentary rocks.
3. Fossils are important tools used to interpret the geologic past, including changes in environmental conditions and changes in life forms.
4. Reefs are massive biochemical limestone structures. They form by the joining of many external skeletons of colonial organisms like corals.
5. Sedimentary rocks contain much information about Earth history, and they contain many economically important materials like coal and oil.
6. Most seafloor sediments contain the remains of organisms that once lived near the sea surface. Their shells slowly settled to the sea floor as the numbers and types of these organisms changed with time and climate.
7. a. graded bedding; b. cross bedding
8. a. gypsum; b. halite
9. d
10. a. True (as they are made by wind); b. True; c. True (biochemical limestone); d. False. If the pebble-sized particles have sharp edges (they are angular), then the rock is breccia; e. True

CHAPTER 7: METAMORPHIC ROCKS

VOCABULARY REVIEW (Chapter 7)
1. a. regional metamorphism; b. contact or thermal metamorphism
2. parent rocks
3. stress
4. shields
5. a. hydrothermal solution; b. hydrothermal
6. aureole
7. Burial
8. shear
9. index minerals
10. a. Foliation; b. and c. (either order): slaty cleavage; rock cleavage
11. a. schistosity; b. schist
12. nonfoliated
13. Porphyroblastic
14. Gneiss
15. Marble
16. Quartzite

17. Slate
18. Migmatites
19. Texture

APPLYING WHAT YOU HAVE LEARNED (Chapter 7)

1. It is a metamorphic texture characterized by the parallel alignment of mineral crystals and structural features.
2. It is a hard, fine-grained, nonfoliated metamorphic rock commonly formed as a result of contact (thermal) metamorphism.
3. Seawater migrates through cracks in hot, newly formed basaltic rocks of ocean ridges, where it is heated. This hot water removes metallic ions from the basalt and carries them back to the cooler seafloor, where they precipitate from black smokers to form metallic ore deposits.
4. a. They are both metamorphic features produced by the foliation process (and are both foliations); b. They are two different kinds of foliations. Slaty cleavage is a fliation characterized by nearly perfect parallel alignment of mineral grains, has been caused by a low grade of metamorphism, and has undergone almost no recrystallization. Schistosity is a foliation characterized by less-parallel alignment of mineral grains, has been caused by a relatively high grade of metamorphism, and has undergone much recrystallization.
5. a. nonfoliated, hard, dense, rock formed from the metamorphism of quartz sandstone; b. fine-grained, foliated, metamorphic rock with nearly perfect rock cleavage (slaty cleavage); c. foliated metamorphic rock with alternations in mineralogy that give the rock a banded appearance; d. fine-grained, foliated metamorphic rock with a metallic sheen and often having a wrinkled appearance; e. coarse-grained, nonfoliated metamorphic rock composed chiefly of calcite and often composed of equidimensional crystals

ACTIVITIES AND PROBLEMS (Chapter 7)

1. They form when partial melting of rocks occurs at very high temperatures. Light-colored minerals then melt to form bands of magma, while the dark-colored minerals remain relatively solid. Migmatite forms if the partially melted rock body slowly cools, and the magma bands crystallize into igneous rock bands within the dark-colored body of metamorphic rock.
2. One would expect zones of high-pressure metamorphism to occur wherever there is much compression of crust or lithosphere. This could occur wherever plates are colliding (at convergent plate boundaries).
3. a. Weathering of preexisting rocks occurred at the earth's surface.
   b. Erosion, transportation, and rounding of weathered rock fragments caused them to develop spherical shapes.
   c. The spherical rock fragments and associated clay, silt, and sand were deposited as a layer of gravel.
   d. Burial of the gravel layer caused it to become lithified into a conglomerate.
   e. Continued burial caused heating, so much that sedimentary grains became very pliable (nonrigid).
   f. Shear and continued burial transformed the conglomerate into metaconglomerate (during which the rock became more dense, shortened in the vertical dimension, and lengthened in the horizontal dimension).
4. Impact metamorphism has occurred when comets and asteroids impacted the earth. It may be that tektites are glass bodies that formed when such impacts melted rocks on impact, and the melted rock particles cooled into glass (tektites).
5. Metamorphism occurs in association with large-scale deformation (regional metamorphism), where rock has been in contact with magma (contact metamorphism), and along fault zones.
6. a. foliated; b. gneiss; c. side to side
7. As solid rocks are subjected to intense heat and pressure, they become less rigid (pliable) and flow in response to stress. When the rock cools, the intricate folds remain. Weathering and erosion of the overburden can allow the intricately folded rocks to be exposed at the surface of the Earth.
8. heat, pressure, chemically active fluids
9. Confining pressure is the pressure on rocks (equal in all directions) at great depth inside the Earth.

10. a. foliated; b. gneiss; c. porphyroblastic and schistosity
11. a. nonfoliated; b. marble

REVIEW EXAM (Chapter 7)
1. The process begins when rock is subjected to conditions unlike those under which it formed. The process ends after the rock undergoes changes to reach a state of equilibrium with the new environmental conditions.
2. The hot water acts as a catalyst to speed up ion migration and cause recrystallization of minerals to larger sizes or to different kinds of minerals. Ions added from the hydrothermal solutions themselves may also change the composition of the rock body as new minerals form from the newly added ions.
3. a. B; b. C; c. A; d. A
4. Hornfels is commonly formed in response to high-temperature, low-pressure contact metamorphism.
5. a. high; b. low; c. high; d. low; e. intermediate; f. high; g. intermediate or high
6. Along ocean ridges and at oceanic hot spots, sea water circulates through still-hot basaltic rocks, transforming the iron- and magnesium-rich minerals into metamorphic minerals such as talc and serpentine.
7. b
8. a. False. It generally encourages the growth of larger crystals of the same mineral(s); b. True; c. True; d. False. The parent rock for marble is limestone or dolostone (dolomite-rich rock). Quartz sandstone is metamorphosed into quartzite; e. False. Gneiss commonly develops its banded texture as a result of foliation; f. False. Contact metamorphic rocks usually have a fine-grained texture.

CHAPTER 8: GEOLOGIC TIME

VOCABULARY REVIEW (Chapter 8)
1. Relative dating
2. Numerical dates
3. law of superposition
4. principle of cross-cutting relationships
5. principle of original horizontality
6. Inclusions
7. a. conformable; b. unconformities
8. a. correlation; b. principle of fossil succession; c. index fossils
9. Angular unconformities
10. Nonconformities
11. Radiocarbon dating
12. a. radioactivity; b. radiometric dating
13. disconformity
14. a. eon; b. eras; c. epochs; d. periods
15. daughter products
16. half-life
17. a. Paleozoic; b. Cenozoic; c. Mesozoic
18. Phanerozoic
19. a. Precambrian; b. Hadean; c. Archean; d. Proterozoic
20. geologic time scale

APPLYING WHAT YOU HAVE LEARNED (Chapter 8)
1. An unconformity is any break in the rock record caused by an interruption of deposition or by removal of previously formed rock.

2. In an undeformed sedimentary sequence of strata, each bed (layer) is older than the one on top and younger than the one below.
3. Layers of sediment (strata) are generally deposited in a nearly horizontal position.
4. When rocks are cut by another feature (e.g., fault, fracture, dike), then they must be older than that feature.
5. It is because fossil plants and animals succeed one another in a definite and determinable order (the principle of fossil succession), which is the basis for subdividing and naming intervals of geologic time.
6. Many fossils are much older than 100 million years (based on radiometric dating of the rocks in which they are found).
   Radiocarbon dating cannot be used to date fossils that are older than 75,000 years.
7. It is the task of matching up rocks of similar ages among different regions.
8. It is the time required for half of a parent isotope to decay to its stable daughter products.
9. a. nonconformity; b. disconformity; c. angular unconformity
10. a. alpha emission (alpha decay); b. 2 fewer; c. 4 fewer; d. beta emission (beta decay); e. no change; f. 1 more; g. electron capture; h. 1 fewer; i. no change

ACTIVITIES AND PROBLEMS (Chapter 8)
1. a. 2; b. 3
2. The sample has undergone 3 half-lives, so 3 x 1.3 billion years is 3.9 billion years.
3. The sample has undergone 2 half-lives.
4. a. 1/256; b. 1/256 = 0.0039, and 0.0039 x 100 equals 0.39%; c. 1/128; d. 1/128 = 0.0078, and 0.0078 × 100 equals 0.78%; e. No, with that amount of error in measuring the remaining amount of parent material, it would not be possible to determine exactly how many half-lives of decay the sample has undergone.
5. It is because the amount of parent material remaining after 75,000 years, and the exact amount of half-lives that have elapsed, cannot be determined accurately.
6. a. 234 − 8 = 226; b. 92 − 4 = 88
7. Radon is a colorless, odorless, invisible gas that is radioactive. It is a radiation hazard and the second worse cause of lung cancer (after smoking). Radon originates in the ground from the radioactive decay of natural uranium. It travels into homes through air and water.
8. a. b; b. f
9. a. h; b. t

REVIEW EXAM (Chapter 8)
1. a. A sequence of horizontal strata is deposited.
   b. The sequence of horizontal strata is folded or tilted, then partly eroded.
   c. A second (younger) sequence of horizontal strata is deposited on the partly eroded and folded/tilted strata.
2. Method a: The total thickness of all sedimentary strata deposited during Earth's history was estimated, then divided by the estimated average rate at which sediments accumulate.
   Problems were that rates of sediment accumulation vary, estimates of the total thickness of strata were not consistent, and no corrections were made for compaction.
   Method b: The total salt content of the oceans was estimated and then divided by the amount of salt added to the oceans each year. Problems were that estimates of ocean volume and salt content were not consistent, the amount of salt added each year varies, and it is possible that salt is periodically removed from the oceans.
3. a. 4.5; b. 4,600
4. angular unconformity
5. Rock containing inclusions is younger than the inclusions.
6. The Precambrian contains few fossils, if any.

7. Relative age relationships determined from fossils and physical relationships of rock units were used to name time and rock subdivisions and place them in order of their relative ages. Much later, numerical dates were added to complete the time scale.
8. Possibilities include:
    a. Ratios of isotopes in a rock unit may be altered by heating of the rock unit.
    b. The material (wood, shell, crystal) containing the isotopes used in the radiometric dating process may have been moved from its original position and relocated in a new one.
9. g
10. b
11. e
12. b
13. a
14. One reason why the granite is older than the sedimentary layers is that the granite occurs beneath the sedimentary layers (i.e., it was there before the sedimentary layers were deposited on top of it). Also, there are inclusions of granite in the oldest of the sedimentary layers, which means that parts of the granite were eroded and incorporated as sedimentary grains into the sedimentary layers.
15. a. Cenozoic; b. Mesozoic; c. Paleozoic
16. a. True; b. True; c. False. It is composed of the Triassic, Jurassic, and Cretaceous Periods; d. True; e. False. He estimated that the earth is about 6,000 years old; f. True; g. False. William Smith made such correlations; h. False. Steno is credited with recognizing these items; i. False. It marks the end of the Mesozoic Era.

CHAPTER 9: MASS WASTING—THE WORK OF GRAVITY

VOCABULARY REVIEW (Chapter 9)
1. mass wasting
2. angle of repose
3. a. fall; b. slide; c. flow
4. rock avalanches
5. creep
6. a. permafrost; b. solifluction
7. slump
8. a. rockslides; b. debris slide
9. a. debris flow; b. lahar

APPLYING WHAT YOU HAVE LEARNED (Chapter 9)
1. Rockslides are movements that occur when blocks of bedrock break loose and slide down a slope, whereas slump is a form of slope failure where the solid rock or unconsolidated material forming the slope slips along curved, spoon-shaped surfaces.
2. a. scarp; b. earthflow (See Figure 9.10 in your textbook.)
3. It is any downslope movement of rock, regolith, or soil under the influence of gravity.
4. A slide is any movement of material along a well-defined surface.
5. A fall is any free-fall of detached pieces of rock, regolith, or soil.
6. A flow is any movement of material as a viscous fluid.

ACTIVITIES AND PROBLEMS (Chapter 9)
1. gravity, water, angle of repose (oversteepened slopes), vegetation, earthquakes
2. It is because little vegetation is present to hold unconsolidated material, so the infrequent, heavy rains create fluid mixtures of mud, soil, rock, and water that flow down canyons and gullies.

3. Debris flows typically flow rapidly down channels and gullies that occur on a slope, whereas earthflows typically flow more slowly and involve teardrop-shaped or tongue-shaped flows of a portion of a slope.
4. In general, stream valleys are produced by mass wasting and the erosional and transporting effects of water running in channels.
5. Erosion requires a transporting medium (wind, water, or ice) in order to occur, whereas mass wastage does not require a transporting medium in order to occur.
6. type of material involved, the kind of motion displayed, and the velocity of the movement
7. Prediction: The cliff will fail and the building will slide/fall into the ocean.
   Explanation: Ocean waves will cut into and erode the base of the cliff, so the cliff will fail and slide into the ocean (along with the building that was supported by the material composing the cliff).
8. Prediction: The building will be buried or destroyed by some form of mass wastage, probably a rockslide.
   Explanation: Rainwater will seep downward through the porous sandstone, but it will not pass through the clay. As a result, the upper surface of the clay will become very lubricated, and one or more blocks of the sandstone will slide down the lubricated surface (and so bury or destroy the building).

REVIEW EXAM (Chapter 9)
1. a. slump; b. rockslide; c. debris flow; d. earthflow (See Figure 9.6 of your textbook.)
2. gravity
3. Water lubricates surfaces, and it adds weight. Both factors increase the likelihood that mass wastage will occur.
4. Permafrost occurs in regions where summertime temperatures do not get high enough for long enough periods to melt more than a shallow surface layer of the frozen ground.
5. Creep is caused by alternating periods of freezing and thawing or wetting and drying. The freezing or drying lifts particles, but the thawing or wetting allows the particles to fall back down and move slightly downslope.
6. e
7. a. False. They also occur on slopes under water; b. False. The process can be rapid, with known movement of up to 200 km per hour; c. True; d. True; e. False. More like 25% of the Earth's surface is underlain by permafrost; f. False. When homes are built on permafrost, they melt the permafrost and the house subsides and cracks; g. True

CHAPTER 10: RUNNING WATER

VOCABULARY REVIEW (Chapter 10)
1. hydrologic cycle
2. a. infiltration; b. runoff
3. a. transpiration; b. evapotranspiration
4. a. sheet flow; b. rills; c. stream
5. infiltration capacity
6. gradient
7. discharge
8. a. laminar flow; b. turbulent flow
9. a. head or headwaters; b. head or headwaters; c. mouth
10. longitudinal profile
11. a. base level; b. local base level; c. ultimate base level
12. graded stream
13. potholes
14. a. dissolved load; b. suspended load; c. bed load

15. settling velocity
16. a. natural levees; b. back swamps; c. yazoo tributaries
17. floodplain
18. Braided
19. saltation
20. a. capacity; b. competence
21. a. Sorting; b. alluvium
22. bars
23. a. Playfair's law; b. drainage basin; c. divide
24. a. Alluvial fans; b. deltas; c. distributaries
25. a. meanders; b. cut banks; c. cutoff; d. oxbow lake; e. meander scar
26. a. water gap; b. wind gap
27. Headward erosion
28. Stream piracy
29. terraces
30. peneplane
31. a. rejuvenated; b. entrenched meanders

APPLYING WHAT YOU HAVE LEARNED (Chapter 10)
1. It is the combined effect of the process of evaporation and the process of transpiration.
2. It is that part of a stream valley that is inundated during floods.
3. It is a stream that flows behind a natural levee of a larger stream, and also parallels the larger stream, until it is able to make its way into the larger stream.
4. It is the entire land area that contributes water to a stream system.
5. Channel deposits are mostly composed of sand and gravel deposited as bars, but the floodplain deposits are gravels, sands, silts, and clays deposited as levees and in back swamps.
6. a. rectangular pattern; b. trellis pattern; c. dendritic pattern; d. radial pattern
7. a. neck; b. point bar; c. cut bank

ACTIVITIES AND PROBLEMS (Chapter 10)
1. a. no; b. As meander cutoff occurs along one segment of the stream, new meanders form along other parts of the stream.
2. As erosion occurs along the outer edge of meanders (along the cut bank), deposition of point bars occurs on the inside edge of meanders.
3. Refer to Box 10.1 in your textbook. Here are four reasons:
   a. Runoff from spring thaw of the southern region joins snowmelt from the north (because the river runs north.)
   b. Ice sometimes blocks water flow in the winter and spring, causing the river to flood behind the ice dams.
   c. The river flows across a broad plain, so a flood travels great distances.
   d. The river's gradient is low.
4. Water from the oceans and standing bodies of water on the land undergo some evaporation as plants release water to the atmosphere by the process of transpiration. The net result (evapotranspiration) is moisture in the atmosphere that eventually forms clouds. Precipitation from the clouds allows the water to fall back to the earth's surface, where some runs back into oceans and water bodies on the land, and some soaks into the ground. Much of the water that falls on, or soaks into, the land eventually makes its way back into the ocean to complete the cycle.
5. a. 2 m/sec x 10 m width x 1 m depth = 20 m^3/second; b. double the stream's velocity to 4 m/sec, or narrow the area of the channel by half to 5 m^2

REVIEW EXAM (Chapter 10)

1. the intensity and duration of the rainfall; the soil texture; the wetness of the soil before the storm; the slope of the land; the nature of the vegetation cover
2. a. Mississippi; b. Amazon
3. c
4. c
5. b
6. c
7. d
8. d
9. b
10. by lifting loose particles; by abrasion; by solution activity
11. a. increases; b. decreases; c. increases; d. increases; e. increases
12. a. gradient; b. cross-sectional shape of the channel; c. size and roughness of the channel; d. the stream's discharge
13. in solution (dissolved load), in suspension (suspended load), and along the bottom (bed load)
14. Rivers remained above flood stage for months throughout summer, multiple flood crests occurred, and flood crests set many record highs.
    a. The river that enters the area from the left (through the water gaps) must have once flowed to the right side of the area, as indicated by the presence of wind gaps.
    b. The area was then structurally tilted (down toward the front of the area) so that the river entered drainage that runs from back to front (top to bottom) of the area, thereby creating the abrupt bend (i.e., stream piracy occurred).
15. a. False. It is powered chiefly by energy from the sun; b. True; c. True

CHAPTER 11: GROUNDWATER

VOCABULARY REVIEW (Chapter 11)

1. porosity
2. permeability
3. a. aquifers; b. aquitards
4. a. gaining; b. losing
5. a. belt of soil moisture; b. capillary fringe; c. zone of aeration
6. a. zone of saturation; b. water table; c. groundwater
7. hydraulic gradient
8. Head
9. perched water table
10. well
11. spring
12. a. drawdown; b. cone of depression
13. a. artesian; b. nonflowing artesian well
14. flowing artesian well
15. a. Geysers; b. geyserite or siliceous sinter; c. geyserite or siliceous sinter; d. calcareous tufa or travertine; e. calcareous tufa or travertine
16. caverns or caves
17. a. speleothems; b. stalactites; c. stalagmites
18. karst topography
19. a. sinkholes or sinks; b. sinkholes or sinks

APPLYING WHAT YOU HAVE LEARNED (Chapter 11)

1. Darcy's law states that if permeability remains uniform, then the velocity of groundwater flow will increase as the slope of the water table increases.
2. It is obtained by dividing the vertical difference between the recharge and discharge points (head) by the length of the flow between those points.
3. It is a conical depression in the water table caused by drawdown around a well.
4. a. water table; b. perched water table
5. a. influent stream; b. effluent stream
6. They are springs having water that is at least 6-9°C warmer than the mean annual air temperature for the localities where they occur. The water is heated underground as it gets near hot igneous rocks.

ACTIVITIES AND PROBLEMS (Chapter 11)

1. a. stalagmite; b. Calcite precipitated from water drops dripping from the stalactite above it; c. stalactite; d. Calcite precipitated from water drops on the cave ceiling before they dripped to the stalagmite below.
2. It is because water is pumped up into a water tower or hilltop reservoir, so that the water pressure surface is above the surface of the land and buildings. Thus, the water tower or reservoir represents the area of recharge, the pipes represent a confined aquifer, and faucets in homes represent flowing artesian wells.
3. Groundwater is confined to an inclined aquifer that is recharged at only one end. The aquifer is confined by aquitards.
4. As water is withdrawn from the ground, the water pressure drops, allowing the land to subside.
5. Use recharge wells to pump waste freshwater back into the groundwater system. Build large basins to collect surface drainage and allow it to seep into the ground.
6. Such aquifers have large, interconnected openings, so contaminated groundwater can travel long distances without being cleansed.
7. a. porous and permeable; b. porous; c. porous; d. porous and permeable

REVIEW EXAM (Chapter 11)

1. a. wells A and C will go dry; b. This is because a cone of depression will form, with its center on the deepest well.
2. intensity of rainfall; steepness of slope; nature of surface material; type and amount of surface vegetation
3. a. aquifer; b. aquifer; c. aquitard; d. aquitard; e. aquifer
4. b
5. e
6. a. Discharge of groundwater exceeds recharge of the acquifer; b. Irrigate less often or use water from other sources.
7. There is not enough water to sustain it into the distant future.
8. a. X; b. There is more open space (porosity) in X. Assuming that this pore space is interconnected, X is a better acquifer than Y.
9. a. True; b. True; c. False. Some do require pumping because the water may not flow out the top of the well even though it flows partly up the well; d. False. The primary erosional work carried out by groundwater is dissolving rock; e. False; f. False. It could start again if groundwater discharge exceeds recharge.

CHAPTER 12: GLACIERS AND GLACIATION

VOCABULARY REVIEW (Chapter 12)

1. glacier
2. a. valley glaciers, alpine glaciers; b. valley glaciers, alpine glaciers; c. Piedmont

3. Ice sheets (or continental glaciers)
4. ice shelves
5. Ice caps
6. outlet glaciers
7. firn
8. a. Basal slip; b. plastic flow
9. a. zone of fracture; b. crevasses
10. surges
11. a. zone of accumulation; b. snowline
12. ablation
13. zone of wastage
14. calving
15. Plucking
16. glacial striations
17. glacial budget
18. a. truncated spurs; b. hanging valleys
19. pater noster lakes
20. a. cirques; b. tarns
21. col
22. Fiords
23. a. arêtes; b. horns
24. Roches moutonnées
25. a. Glacial drift; b. stratified drift, c. till
26. erratics
27. a. end moraine or terminal moraine; b. end moraine or terminal moraine; c. recessional moraines
28. ground moraine
29. lateral moraines
30. Medial moraines
31. a. drumlins; b. karnes
32. Ice-contact deposits
33. kame terraces
34. a. valley train; b. outwash plains; c. kettles (or kettle holes)

APPLYING WHAT YOU HAVE LEARNED (Chapter 12)
1. It is an ice sheet that covers a large portion of a continent.
2. They are boulders found on bedrock surfaces or in till, and their composition is different from the bedrock beneath them.
3. Cirques are bowl-shaped depressions at the heads of glaciated valleys.
4. The terrain of regions eroded by ice sheets is subdued, but the terrain of regions eroded by alpine glaciers is rugged.
5. Antarctic ice sheet, Greenland ice sheet
6. a. truncated spurs; b. arete; c. horne; d. cirque; e. medial moraine
7. a. River alleys are V-shaped, but glaciated valleys are U-shaped.;
   b. A river erodes only the part of the valley that is beneath its narrow channel, whereas a glacier erodes the valley in all areas where the broad mass of ice exists (commonly on the bottom and sides of the valley).
8. a. hanging valleys; b. arête; c. horn; d. tarn; e. pater noster lakes; f. cirques
9. a. Surges are times when glaciers move more rapidly than normal; b. Surges may hold clues about human inpacts on the environment or about climate change.
10. Agassiz and others realized that certain features on Earth's surface can only be formed by glaciers, so they were able to map the extent of now-extinct ice sheets.
11. a. esker; b. drumlin; c. end moraine; d. ground moraine or till; e. kettle hole/lake; f. outwash plain

ACTIVITIES AND PROBLEMS (Chapter 12)

1. c
2. They are moraines marking positions where the ice front of a glacier occasionally stabilized during retreat.
3. a. rate of ice movement
   b. thickness of ice
   c. shape, hardness, and abundance of rock fragments in the ice
   d. erodibility of the surface beneath the ice
4. a. Snow accumulates.
   b. Snow recrystallizes to ice granules called firn.
   c. Compaction and recrystallization of firn forms a solid mass of glacial ice.
5. The glacial theory proposed by Agassiz and others was based upon an understanding of features and deposits created by modem glaciers. Recognizing that certain features are produced by no other known process but glacial activity, they were able to note where vanished glaciers once existed.
6. a. hanging valley; b. A valley feeding into a valley glacier was truncated by erosion of the valley glacier to form a hanging valley.

REVIEW EXAM (Chapter 12)

1. (Any three of the following)
   a. sea level changes
   b. changes of drainage systems
   c. migration of organisms
   d. downwarping and subsequent rebound of the Earth's crust
   e. climate changes
2. Ice ages may occur when shifting lithospheric plates carry land areas to positions near the poles and when mountains develop at some plate boundaries (to provide high altitudes where glaciers form).
3. Variations in Earth's orbit cause variations in the amount of solar radiation received by the Earth, so climate changes (including ice ages) could be caused by such variations in orbit.
4. plucking and abrasion
5. Valley glaciers cut deep valleys into lowland areas during glacial episodes. As the glacial episodes end (interglacials start), sea level rises and the ocean inundates seaward parts of the deep valleys to form fiords.
6. Glaciers cover about 10% of the Earth's land today, but glaciers covered about 30% (or one-third) of the land during the last Pleistocene glaciation.
7. Glacial ice contains trapped air bubbles, atmospheric dust, pollen, and chemical pollution from each of the times when its snow and ice accumulated. These things are clues about how climate has changed.
8. c
9. c
10. d
11. a
12. c
13. a. False. See Figure 12.31 on page 318 of your text; b. True; c. True

CHAPTER 13: DESERTS AND WINDS

VOCABULARY REVIEW (Chapter 13)

1. Xerophytes
2. a. deserts; b. steppes
3. a. equatorial low; b. subtropical highs
4. rainshadow deserts

5. a. deflation; b. blowouts; c. desert pavement
6. a. abrasion; b. ventifacts; c. yardang
7. a. dunes; b. slip face; c. cross beds
8. a. barchans (barchan dunes); b. parabolic dunes
9. a. transverse dunes; b. barchanoid dunes; c. longitudinal dunes
10. ephemeral
11. Loess
12. Interior drainage
13. a. alluvial fan; b. bajada
14. a. playa lakes; b. playas
15. pediment
16. inselberg
17. desertification

APPLYING WHAT YOU HAVE LEARNED (Chapter 13)
1. depressions caused by deflation
2. stones shaped by natural sandblasting
3. a. longitudinal dunes; b. parabolic dunes; c. star dunes; d. barchan dunes; e. transverse dunes; f. barchanoid dunes

ACTIVITIES AND PROBLEMS (Chapter 13)
1. the rainshadow-desert effect; position in deep interior of large landmass; regions of descending air associated with high-pressure zones
2. Wind carries sediment (suspended load); wind rolls and pushes sediment along surfaces (bed load); and wind bounces sediment along surfaces (saltation load).
3. Humid areas are affected chiefly by intense chemical weathering because of the abundance of plants that contribute organic acids and the abundance of water to carry acids and dissolved materials. Plant cover also tends to reduce erosional affects of rainstorms and flooding. Arid areas lack significant plant cover and water, so chemical weathering is reduced. Infrequent rains and the lack of plant cover cause severe erosion and flooding, so mechanical weathering is the chief factor in wearing away the land.
4. (Any three of the following)
   a. wind direction and velocity
   b. availability of sand
   c. amount of vegetation
   d. relief of the surface on which the dune develops
5. Home B would be the safe purchase; b. because the sand dunes are moving from left to right. See Figure 13.15 in your textbook.
6. Home Y would be the safe purchase; b. Sand is being scoured from the blowout and blown toward home X, which will eventually be covered by sand.
7. a. great amount of relief, plateaus developed, pediments, large playa lakes/playas, small alluvial fans, sediment-filled basins poorly developed
   b. reduced relief, no plateaus, playa lakes/playas, large alluvial fans and bajadas, sediment-filled basins moderately well developed
   c. relief greatly reduced to small hills and inselbergs, large bajadas, playa lakes/playas, sediment-filled basins extensively developed
8. As air rises and cools on the windward side of a mountain, precipitation occurs. Therefore, air reaching the leeward side of the mountain is cool and dry. As the cool air descends and warms, it accepts moisture and makes precipitation unlikely.
9. (Any two of the following)
   a. They consist mostly of sand dunes.
   b. They contain practically no life.
   c. Their landforms are shaped chiefly by wind.

10. a. The feature is desert pavement. b. It formed as silt and sand were blown away (by deflation). See Figure 13.12 in your textbook.

REVIEW EXAM (Chapter 13)
1. a. playa or playa lake; b. alluvial fan; c. normal faults bounding a fault block; d. middle stage; e. mountains dissected into an intricate series of valleys and sharp divides, alluvial fans are discrete, pediment moderately developed
2. A climate in which the amount of annual precipitation is not as great as the amount of annual loss of water by evaporation
3. Examples (pick any two) include overgrazing of vegetation by livestock, clearing of land for farming, excessive withdrawal of groundwater, clearing of land for wood to burn for fuel.
4. a. Plate movements can carry landmasses to air-pressure belts where air is descending (high-pressure belts).
   b. Plate movements can carry landmasses to high-temperature, high-evaporation latitudes.
   c. Mountains formed at plate boundaries may create rainshadow deserts.
   d. Plate collisions may create huge landmasses with central deserts that formed because such central areas were so far removed from any source of moisture.
5. It is because temperature and evaporation are also important (as temperature rises, potential evaporation increases); 25 cm of rain per year in Nevada may support little vegetation, but 25 cm of rain per year in northern Scandinavia supports the growth of forests.
6. c
7. a
8. d
9. b
10. c
11. a. True; b. False. The main agent of erosion in deserts is water; c. False. Sand dunes often form only small portions of many deserts. The most common feature of deserts is the lack of significant vegetation; d. False. The greatest depth to which a blowout can develop is to the local water table; e. True

CHAPTER 14: SHORELINES

VOCABULARY REVIEW (Chapter 14)
1. a. wave height; b. wave length
2. Fetch
3. whitecaps
4. a. waves of oscillation; b. waves of translation
5. surf
6. a. swash; b. backwash
7. wave refraction
8. a. beach drift; b. longshore currents
9. a. wave-cut cliffs; b. wave-cut platforms
10. sea arch
11. sea stacks
12. spits
13. a. groin; b. seawall
14. barrier islands
15. tombolo
16. baymouth bar
17. Jetties
18. breakwaters

19. Beach nourishment
20. a. Submergent coasts; b. emergent coasts
21. estuaries
22. a. Tides; b. flood current; c. ebb current; d. tidal flats; e. tidal currents
23. Spring tides
24. Neap tides
25. tidal delta

APPLYING WHAT YOU HAVE LEARNED (Chapter 14)
1. It is a coastline created when sea level falls or the land adjacent to the sea rises.
2. Flood tides occur when tides are rising and inundating (flooding) coastal areas.
3. Beach drift is the tendency for sedimentary grains to zigzag along the beach.
4. A tombolo is a narrow ridge of sand connecting an island to the mainland or to another island.
5. a. baymouth bar; b. wave-cut cliff; c. tombolo; d. spit
6. a. a sandbar that completely crosses the mouth of a bay
   b. isolated pillar or knob of an eroded headland that protrudes above shallow coastal waters
   c. a low sandy island that is elongated parallel to shore and is separated from the mainland by a lagoon
   d. a wall constructed perpendicular to shore for the purpose of trapping sand

ACTIVITIES AND PROBLEMS (Chapter 14)
1. West-coast shorelines are rugged and tectonically active, whereas east-coast shorelines are very subdued, tectonically inactive, and have well-developed coastal plains associated with them.
2. a. Sediment-trapping dams on rivers prevent sand from reaching beaches.
   b. Coastal development increases runoff (and erosion) and adds water to slopes (causing mass wastage).
3. It is because rising sea level is a very slow, gradual process. Shoreline erosion problems are more easily blamed on catastrophic events such as storm effects.
4. a. Relocate buildings away from beaches.
   b. Add sand to replenish beaches (beach nourishment).
   c. Build structures such as seaways to hold the shoreline in place.
5. wind speed; fetch (distance the wind has traveled across open water); length of time the wind has blown
6. As a wave reaches shallow water and touches bottom, the wave slows and translates. As the wave's length becomes shorter, the wave's height grows higher until the wave collapses (breaks).
7. It is the tendency for waves to bend in map view, so that they become nearly parallel to shoreline.
8. Irregular shorelines are attacked by waves from all sides, so the irregularities are eroded, and the coastline becomes straight.
9. wind damage, storm surge, and inland freshwater (rainwater) flooding.

REVIEW EXAM (Chapter 14)
1. Tides result from the gravitational attraction exerted upon the Earth by the moon and, to a lesser extent, by the sun.
2. a. groins; b. breakwaters; c. jetties; d. It will widen as more sand accumulates; e. It will widen as more sand accumulates; f. It will narrow as sand is removed; g. It will widen as more sand accumulates.
3. a. by severing spits from the mainland
   b. as tides heap up sediments to form an offshore island that is elongated parallel to shore
   c. as a coastline with ridges parallel to shoreline is inundated by sea level rise or lowering of the land

4. Carbon dioxide is a heat-absorbing gas, so increasing it should cause higher atmospheric temperatures. This, in turn, will melt glaciers and the increased runoff will cause sea level to rise. The increased temperature will also raise surface temperatures of the ocean, so the water will expand and sea level will rise.
5. a. wave crest; b. wave trough; c. wavelength
6. c
7. d
8. a
9. b
10. a
11. a
12. a. True; b. True; c. False. Loss of life was reduced because of timely storm warnings.

CHAPTER 15: CRUSTAL DEFORMATION

VOCABULARY REVIEW (Chapter 15)
1. Deformation
2. force
3. a. brittle failure; b. ductile
4. a. Stress; b. strain
5. a. elastically; b. ductile deformation
6. geologic structures
7. a. compressional stress; b. tensional stress; c. differential stress
8. outcrops
9. a. Strike; b. Dip
10. Folds
11. a. limbs; b. axial plane; c. axis
12. plunging
13. a. anticline; b. syncline
14. a. symmetrical; b. asymmetrical
15. overturned
16. recumbent
17. monoclines
18. a. domes; b. basins
19. Hogbacks
20. a. Faults; b. fault zone
21. a. Fault gouge; b. slickensides
22. a. dip-slip faults; b. headwall; c. footwall
23. strike-slip faults
24. Fault scarps
25. klippe
26. a. normal faults; b. reverse faults; c. thrust faults
27. detachment
28. fault-block mountains
29. a. horsts; b. grabens
30. right-lateral
31. transform faults
32. joints

APPLYING WHAT YOU HAVE LEARNED (Chapter 15)
1. a. an upfold having the oldest rocks in its center; b. a fold that is "lying on its side"; c. a broad, blister-like upwarp on the crust; d. a fold in which the axis is not horizontal (is inclined)
2. a. normal (dip-slip) fault; b. reverse (dip-slip) fault; c. monocline developed over a normal (dip-slip) fault
3. a. graben; b. horst; c. normal faults
4. a. normal fault; b. The hanging wall has moved down relative to the foot wall.
5. a. recumbent fold; b. The layers have been folded, and the fold is lying on its side.
6. a. axial plane; b. axis; c. limbs; d. anticline

ACTIVITIES AND PROBLEMS (Chapter 15)
1. It is a transform fault (strike-slip fault) with right-lateral strike-slip motion.
2. a. compressional stresses; b. tensional stresses; c. compressional stresses; d. differential stresses; e. compressional stresses
3. Such shapes are determined from the orientation data (strike and dip information, etc.) obtained from many different outcrops.
4. Rocks composed of strong materials (minerals) undergo brittle deformation (granite, gneiss, quartzite); rocks composed of weak materials deform by ductile flow (rock salt, marble, gypsum, shale).
5. Rocks deform when they are subjected to stresses that are greater than their own strength.

REVIEW EXAM (Chapter 15)
1. A dome is a blister-like upwarp of the crust, whereas a basin is a dimple-like downwarp of the crust.
2. When confining pressure and temperature are low, rocks tend to fracture when deformed (i.e., are brittle), but when confining pressure is very high, rocks flow when a force is applied (i.e., they are ductile).
3. a. fold (anticline); b. The layers have been folded into an upfold.
4. the environment (pressure and temperature); strength of the material to be deformed; length of time that the stress is to be applied
5. Strike is the direction of the line produced by the intersection that an inclined rock layer makes with an imaginary horizontal plane.
6. At high confining pressures and temperatures, rocks deform plastically once their elastic limit is surpassed.
7. b
8. a. dome; b. basin
9. a. True; b. False. They are upfolds that have the oldest rocks in the center of the structure; c. False. They are an eroded dome; d. True.

CHAPTER 16: EARTHQUAKES

VOCABULARY REVIEW (Chapter 16)
1. a. earthquake; b. focus; c. epicenter
2. elastic rebound
3. a. foreshocks; b. aftershocks
4. seismology
5. a. seismographs; b. seismogram
6. surface waves or long (L) waves
7. a. primary (P) waves; b. secondary (S) waves
8. a. shallow; b. intermediate; c. deep
9. a. Wadati-Benioff zone; b. 700
10. Modified Mercalli Intensity Scale

11. a. intensity; b. magnitude
12. Richter
13. seismic sea waves or tsunamis
14. liquefaction
15. fault creep
16. seismic gaps
17. moment magnitude

APPLYING WHAT YOU HAVE LEARNED (Chapter 16)

1. a. P-waves, or primary waves, are the fastest seismic waves, which compress and dilate rocks in the direction that they travel. P-waves also travel through solids and liquids, and they are a type of body wave.

   S-waves, or secondary waves, are another type of body wave; they shake particles at right angles to the direction they travel. S-waves do not travel through liquids.

   L-waves, or long waves, are surface waves. They have very large wavelengths and wave heights, so they travel more slowly than the P-waves or S-waves; however, they travel through solids and liquids.

   b. long (L) waves
2. a. C or D; b. D; c. G; d. B or A; e. B or A
3. It is a zone in which earthquakes occur, extending from a trench to as deep as 700 km within the earth. The earthquakes indicate a subduction zone.
4. Earthquake magnitude is a measure of the strength of an earthquake (amount of energy released during an earthquake).
5. Earthquake intensity is a measure of the effects of an earthquake at a particular location.
6. Richter magnitude is calculated by measuring the amplitude of the largest seismic wave recorded, but moment magnitude is calculated using several factors (including average displacement of the fault, area of fault displacement, and sheer strength of the faulted rock). Moment magnitudes have been calibrated so that small and moderate earthquakes have moment magnitude values equivalent to Richter magnitudes. However, only moment magnitudes (not Richter magnitudes) are adequate for representing the strength of large earthquakes.

ACTIVITIES AND PROBLEMS (Chapter 16)

1. Tectonic forces slowly deform crustal rocks on both sides of a fault, so the rocks bend and store elastic energy. When forces holding the rocks together (friction along the fault) are overcome, slippage (an earthquake) occurs, and the rocks spring back to nearly their same position before deformation (i.e., they undergo elastic rebound).
2. They have a mass suspended from a support that is attached to the ground. When an earthquake vibration reaches the instrument, inertia of the mass keeps it stationary while the earth and support move. The movement of the earth (support) in relation to the stationary mass is recorded on a rotating drum or magnetic tape.
3. A tenfold increase in wave amplitude corresponds to an increase of one on the Richter magnitude scale.
4. $10 \times 10 = 100$ times greater
5. The Richter magnitude is 3.0 magnitude units greater ($6.3 - 3.3 = 3.0$), so $30 \times 30 \times 30 = 27,000$ times greater energy was released.

REVIEW EXAM (Chapter 16)

1. c
2. c
3. a
4. b
5. c

6. d
7. the intensity and duration of the vibrations; the nature of the material upon which the buildings rest; the design of the buildings
8. a. C; b. Many earthquakes have occurred along faults at B, and especially at A, so fault creep has occurred there during the small earthquakes. As no small earthquakes have occurred along the fault at C, the elastic energy is still stored there along a "locked" part of the major fault. When the friction along the major fault at C is overcome by the elastic energy built up in the rocks there, a large earthquake may occur.
9. Buildings constructed on loose material (gravel, sand, clay) sustain far more damage than buildings constructed on rigid material (dense, solid bedrock)
10. Most earthquakes occur in narrow belts within the Earth's crust. These belts correspond to boundaries between lithospheric plates, where there is much motion. Few earthquakes occur within the lithospheric plates.
11. Seismic sea waves in the open ocean generally go undetected, because they cause wave swells of about a meter or less. In shallow water, however, seismic sea waves pile up to heights of 30 m or more and crash onto shore.
12. a. home Y; b. Loose sand under home Y would experience liquefaction during a moderate or strong earthquake, so home Y would settle down into the sand or fall over (because the sand grains would vibrate apart in an earthquake).
13. a. True; b. True; c. False.; d. False. A few occur within the plates, even though most occur along the plate boundaries; e. True; f. False. See Box 16.1 in your textbook; g. True

CHAPTER 17: EARTH'S INTERIOR

VOCABULARY REVIEW (Chapter 17)
1. discontinuity
2. a. crust; b. mantle; c. core
3. P-wave shadow zone
4. a. Mohorovicic discontinuity or Moho; b. Lehmann discontinuity
5. a. asthenosphere; b. lithosphere
6. a. mesosphere; b. D layer
7. geothermal gradient
8. conduction
9. convection

APPLYING WHAT YOU HAVE LEARNED (Chapter 17)
1. It is the belt where direct seismic P-waves are absent because they are refracted (bent) into the core.
2. It is a weak (easily deformed), partly molten zone of rock located from about 100 to 700 km deep within the Earth.
3. It is the outermost layer of the Earth that is cool enough to behave like a brittle solid. It is about 100 km thick, includes the crust and uppermost mantle, and is underlain by the asthenosphere.
4. The crust is a very thin, rocky, outer layer of the Earth, less than 35 km thick and composed of rocks of granitic through basaltic composition.
5. a. base of lithosphere or top of asthenosphere (top of low velocity zone); b. inner core; c. lower mantle (mesosphere); d. asthenosphere

ACTIVITIES AND PROBLEMS (Chapter 17)
1. Seismic tomography is way of making images of Earth's interior by using a computer to combine earthquake data from multiple sources to build a three-dimensional image. The images reveal regions of Earth's interior where seismic wave velocities are slow (indicating the presence of hot, upwelling rock) or fast (indicating the presence of cool descending rock).

2. It is probably generated from the flow of molten iron alloy(s) in Earth's outer core. The flow may becaused by a combination of Earth's rotation and the unequal heating of the core. Mechanical energy associated with the flowing iron is converted into magnetic energy (as a self-sustaining dynamo).

3. Meteorites are assumed to represent material like that from which the Earth originally accreted. As meteorites range from iron-nickel metallic types to stony types resembling peridotite, it is likely that these materials form some of the layers of Earth's interior.

4. a. Velocity of seismic waves depends partly on the density and elasticity of materials they pass through.
   b. Seismic wave velocities tend to increase with depth inside the earth.
   c. P-waves travel through solids and liquids, but S-waves travel through solids only.
   d. In all materials, P-waves travel faster than S-waves.
   e. Seismic waves are bent (refracted) as they pass from one material to another.

5. S-waves cannot pass through it, and P-waves decrease 40% as they encounter it because of the reduced elasticity of the material there.

REVIEW EXAM (Chapter 17)
1. Crust: thin (less than 35 km), solid, outer layer composed of granitic through basaltic rocks
   Mantle: 2,885-km-thick rocky layer beneath the crust having the composition of peridotite (at least in its upper part)
   Core: 3,486-km-thick layer beneath the mantle; composed chiefly of iron-nickel alloy(s) and some carbon and silicon

2. He presented the first convincing evidence for layering within the Earth by defining the discontinuity between the crust and the mantle (now called the Mohorovicic discontinuity, or Moho).

3. The shadow zone could be explained if the Earth contained a core composed of material unlike the overlying mantle, so that waves are bent at the discontinuity between mantle and core.

4. a. B; b. A; c. C
5. d
6. b
7. a
8. c
9. d
10. a. False. Fragments of rock from the mantle can be found in kimberlites exposed at the surface of the earth; b. False. Earth's core is probably composed of an iron-nickel alloy; c. True; d. False. The inner core is solid and the outer core behaves as a liquid.

CHAPTER 18: THE OCEAN FLOOR

VOCABULARY REVIEW (Chapter 18)
1. echo sounder
2. a. submarine canyons; b. turbidity flows; c. turbidites; d. deep-sea fan; e. continental rise
3. graded bedding
4. a. continental margin; b. continental shelf; c. continental slope; d. continental rise
5. a. deep-ocean basin; b. abyssal plains
6. trenches
7. a. seamounts; b. guyots or tablemounts
8. atoll
9. a. Terrigenous sediments; b. Biogenous sediments; c. Hydrogenous sediments
10. manganese nodules
11. a. mid-ocean ridges; b. rift zones; c. rift valleys
12. a. pillow; b. sheeted dikes; c. ophiolite complex

13. accretionary wedge
14. coral reefs

APPLYING WHAT YOU HAVE LEARNED (Chapter 18)
1. It is the deposit formed as sediments accumulate after passage of a turbidity current. Part of the deposit is a graded bed.
2. An abyssal plain is a broad, level area of the deep-ocean basin.
3. They are isolated volcanic peaks that occur on the ocean floor.
4. It is sediment consisting chiefly of mineral grains weathered from continental rocks.
5. A passive continental margin generally has a well-developed continental shelf, slope, and rise. However, an active continental margin generally has a well-developed accretionary wedge.

ACTIVITIES AND PROBLEMS (Chapter 18)
1. Turbidity currents are dense currents that flow along the bottom of the ocean. As they flow over the continental slope, they erode submarine canyons.
2. abyssal plains: broad flat areas; seamounts: isolated steep-sided volcanic peaks; deep ocean trenches: deep linear troughs in the sea floor
3. a. Continental shelf: the gently sloping, submerged surface that extends seaward from the shoreline
   b. Continental slope: steeply inclined surface at the seaward edge of the continental shelf
   c. Continental rise: gently inclined surface at the base of the continental slope.
4. Calcareous ooze is mud made up of the calcareous shells of microscopic organisms, which inhabit warm surface waters of the ocean. When the calcareous shells sink into deep cold water, they begin to dissolve (because the water is rich in carbon dioxide, making it acidic). Most of the shells dissolve completely before they can settle onto the floors of deep-ocean basins.

REVIEW EXAM (Chapter 18)
1. Coral reefs form on the flanks of sinking volcanic islands. As the islands sink, the corals continuously grow upward to sea level.
2. a. Terrigenous sediments: sediments composed chiefly of mineral grains weathered from the continents and transported into the oceans.
   b. Biogenous sediments: sediments composed chiefly of the shells and skeletons of marine plants and animals.
   c. Hydrogenous sediments: sediments composed chiefly of mineral grains that crystallized directly from seawater via various chemical reactions.
3. Partial melting of the asthenosphere produces magma of basaltic (gabbroic) composition. The magma that reaches the seafloor produces extrusive pillow basalts. The magma gets to the seafloor via numerous fractures and produces sheeted dikes. The magma that does not flow through the sheeted dikes cools to form gabbro.
4. Newly formed oceanic crust is warmer than other seafloor basalt, so it is less dense and occupies more volume than the cool basalt. This causes the oceanridge basalts to sit higher than the basalts on the deep-ocean floor. The rate of spreading along the Mid-Atlantic Ridge is less than along the East Pacific Rise, so the warm rocks have more time to move upward and form more elevated mountains.
5. The steps are:
   a. Upwelling magma forms a dome.
   b. The dome fractures into a three-armed pattern.
   c. One of the arms fails to develop any more. The other two arms develop into a new ocean basin.
6. a. continental shelf; b. continental slope; c. continental rise
7. c
8. a
9. c

10. a. False. Seafloor spreading carries the atolls into deeper water; b. True; c. True; d. False. They are a type of hydrogenous sediment.

CHAPTER 19: PLATE TECTONICS

VOCABULARY REVIEW (Chapter 19)
1. a. Pangaea; b. continental drift
2. a. Curie points; b. paleomagnetism
3. seafloor spreading
4. a. normal polarity; b. reversed polarity
5. magnetometers
6. plates
7. a. convergent plate boundaries; b. deep-ocean trench; c. accretionary wedge
8. divergent plate boundaries
9. Transform fault boundaries
10. rifts or rift valleys
11. hot spots
12. a. subduction zone; b. deep-ocean trench
13. a. continental volcanic arcs; b. island volcanic arc
14. Fracture zones
15. a. lithosphere; b. asthenosphere
16. a. slab-pull; b. Ridge-push

APPLYING WHAT YOU HAVE LEARNED (Chapter 19)
1. a. divergent plate boundary; b. convergent plate boundary; c. transform fault boundary
2. It is the temperature at which cooling iron-rich minerals become magnetized in the direction parallel to the existing magnetic field.
3. Reversed polarity is a condition of the Earth's magnetic field when it exhibits a polarity that is exactly the opposite of the present polarity of the Earth's magnetic field.
4. A rift is a down-faulted valley or system of valleys that develops where two plates are diverging.
5. a. trench (deep-sea trench); b. ridge (ridge crest); c. transform fault

ACTIVITIES AND PROBLEMS (Chapter 19)
1. Lithosphere is the rigid layer of crust and upper mantle. The lithosphere rests on the asthenosphere, a weaker layer of the mantle. According to the theory of plate tectonics, the lithosphere is broken into numerous pieces, called lithospheric plates, that are moved about and interact because of mantle convection.
2. As new seafloor is created at the ridge crests, it records the polarity of the magnetic field at the time it formed. In this way, each reversal of the Earth's magnetic field polarity is recorded in rocks (basalts) of the seafloor almost as if the seafloor were a giant tape recorder.
3. Oceanic-oceanic convergence results in subduction of one of the plates and development of an arc-shaped line of volcanic islands called a volcanic island arc. Oceanic-continental convergence results in subduction of the leading edge of the oceanic plate and formation of a continental volcanic arc. Continental-continental convergence results in a plate collision lacking subduction, so a mountain belt forms.
4. First, the continents could be fitted together like pieces of a giant jigsaw puzzle. Once the continents were fitted back together, however, patterns created by the distribution of plants and animals, paleoclimate belts, geologic rock types, and structural features all matched up as if painted on the pieces of the puzzle.

5. Hess proposed a seafloor spreading hypothesis to explain continental drift, where ocean ridges are sites of magma upwelling (to create crust) and trenches are associated with crustal subduction (to destroy crust). The seafloors move like a giant conveyor belt.

6. The Earth's magnetic field periodically reverses its polarity, so that the north magnetic pole becomes the south magnetic pole and vice versa. As new basalt is added to the seafloor on both sides of an ocean ridge, it is magnetized according to the magnetic field at that specific time. This causes symmetrical pairs of magnetic polarity stripes to develop parallel to one another but on opposite sides of the ridge. The existence of such magnetic polarity stripes is evidence in favor of the seafloor spreading hypothesis.

7. rafting (floating across the ocean); island hopping (jumping from one island stepping stone to another); and land bridges (walking across temporarily developed land bridges that no longer exist)

8. One model proposes that mantle convection occurs in two layers separated at a depth of 660 km, but seismic imaging (tomography) shows that subducting crust goes below 660 km. A second model proposes that a single layer of convection stirs the entire thickness of the mantle, but whole-mantle mixing would have already eliminated the kind of magma observed in hot spots. The third model proposes that the bottom third of the mantle bulges upward in some areas and sinks in others, without much mixing, but there is little evidence of such a bottom layer in seismic imaging studies (tomography).

REVIEW EXAM (Chapter 19)

1. a. Convergent plate boundaries: plates move toward each other in exactly opposite directions.
   b. Divergent plate boundaries: plates move away from each other in opposite directions.
   c. Transform fault boundaries: plates slide past one another in opposite directions.

2. Transform faults provide the means by which ocean crust created at ridge crests can move to sites of plate destruction (subduction zones). The relative motion of plates can be changed (transformed) along such faults. For example, where transform faults offset ridge segments, they permit the ridge to change direction (curve).

3. This is an example of an oceanic-oceanic convergent plate boundary. One plate is subducting beneath another to form a volcanic island arc.

4. a. Rising magma upwarped the continental crust, causing numerous cracks to develop in the rigid lithosphere.
   b. As the crust is pulled apart, large slabs of lithosphere sink in response to the pull of gravity.
   c. The lakes will become oceans because as the rift widens, ocean water will fill in the low areas to make a narrow sea and eventually a broad ocean.

5. Deep-focus earthquakes can be generated in a manner similar to shallow ones, through the release of elastic energy stored in a descending slab of seafloor rock as it meets resistance in the subduction process. This means that the cool slab of brittle rock descends into the hot asthenosphere some 700 Ian before it is heated enough to now rather than break as a brittle solid. The presence of the deep-focus earthquakes means that cool rocks are being pushed or pulled into the mantle, where they eventually melt, as predicted by the plate tectonics theory.

6. a. The oldest (deepest) sediment from each core can be plotted against its distance from the ridge crest, whereupon it is revealed that the age of such sediments increases with increasing distance from the ridge crest.
   b. No sediments older than about 160 million years old were found, indicating that ocean basins are geologically youthful.
   c. Sediments are thin on the ridge crests and thicken away from the ridge crests.

7. a. Elevated lithosphere may slide under the influence of gravity and push its way toward continents and subduction zones (ridge-push hypothesis).
   b. Subducting plates may pull oceanic materials downward into subduction zones (slab-pull hypothesis).
   c. Convection currents in the mantle may drag lithosphere in motion with them (convection current hypothesis)
   d. Hot spots may have significant enough convection of magma to push plates.

8. d
9. d
10. c
11. a
12. a. False; b. True; c. False. For example, California's San Andreas fault is a transform fault developed on land; d. False. They occur in subduction zones but not below ridge crests.

CHAPTER 20: MOUNTAIN BUILDING AND THE EVOLUTION OF CONTINENTS

VOCABULARY REVIEW (Chapter 20)
1. orogenesis
2. mountain ranges
3. a. isostasy; b. isostatic adjustment
4. accretionary wedge
5. passive margin
6. Terranes
7. fault-block mountains

APPLYING WHAT YOU HAVE LEARNED (Chapter 20)
1. Isostasy is the concept that crust floats on the mantle in gravitational balance.
2. Isostatic adjustment occurs whenever the thickness of crust in a region changes and a new level of gravitational equilibrium is achieved. It occurs when the crust is thickened (e.g., by continental collision) or thinned (e.g., by weathering and erosion).
3. An accretionary wedge is a deposit of sediments that is scraped off a subducting plate and piles up in front of the overriding plate.
4. Orogenesis is the name for all of the processes that collectively produce a mountain system.

ACTIVITIES AND PROBLEMS (Chapter 20)
1. It is because the Piedmont is composed of deformed metamorphic rocks, rocks that have undergone orogenesis.
2. a. No; b. It is because the rocks are still undeformed. They have not undergone orogenesis.
3. isostasy, strength of the lithosphere, and buoyancy of hot rising magma
4. They form where two oceanic plates converge, and one subducts beneath the other. Partial melting of the subducted plate generates magma that migrates upward to form volcanoes of the island arc.
5. India split from Antarctica and drifted north until it collided with Asia. The combined thickness of two continental crusts and oceanic sediments squeezed between them has resulted in development of the massive, tall Himalayan Mountains.
6. The Alps are the result of a collision between Africa and Europe during the closing of the Tethys Sea. Large quantities of seafloor sediments from the ocean between Africa and Europe were squeezed into huge recumbent folds that were thrust-faulted northward.

REVIEW EXAM (Chapter 20)
1. He suggested that the Earth's crustal rocks float on the denser, more plastic mantle and that the crust must be thicker under mountains than beneath lowlands (i.e., mountains have roots).
2. As an oceanic plate subducts beneath a continental margin, partial melting of the subducted plate occurs. Partial melting generates magma that rises along the continental margin to form volcanoes of the volcanic arc.

3. As oceanic plates with embedded island arcs or microcontinents move toward a subduction zone, upper parts of these plates with thickened crust are peeled off and thrust upon the adjacent continental block in relatively thin sheets.
4. 3
5. The upward convective flow of mantle superplumes may explain the elevation of southern Africa, the Colorado Plateau, and the Basin and Range Province. The downward convective flow of mantle superplumes may produce large basins.
6. One view is that most continental crust formed early in Earth's history and has since been modified by tectonic activity. The opposing view is that the original amount of crust was small early in Earth's history, and the continents have grown in size through time by the accretion of material derived from the upper mantle.
7. a
8. a. True; b. True; c. True; d. True; e. False. Periods of mountain building generally last for tens of millions of years to even more than 100 million years.

CHAPTER 21: ENERGY AND MINERAL RESOURCES

VOCABULARY REVIEW (Chapter 21)
1. a. renewable; b. nonrenewable
2. fossil fuels
3. a. Mineral resources; b. metallic mineral resources; c. nonmetallic mineral resources
4. a. oil trap; b. reservoir rock; c. cap rock
5. nuclear fission
6. Reserves
7. Ore
8. Pegmatites
9. hydrothermal fluids
10. a. vein deposits (veins); b. disseminated deposits
11. Secondary enrichment
12. Placers
13. hydroelectric

APPLYING WHAT YOU HAVE LEARNED (Chapter 21)
1. It is a resource that is replaced so slowly that significant deposits of it take millions of years to form.
2. It is a geological environment that allows economically significant amounts of oil and/or gas to accumulate.
3. It is a metallic mineral that can be mined at a profit.
4. It is the process whereby weathering concentrates minor amounts of metals that were scattered throughout unweathered rock, so that an ore is produced.
5. A placer is an ore body formed by current sorting and concentration of heavy minerals.

ACTIVITIES AND PROBLEMS (Chapter 21)
1. coal, petroleum (oil), natural gas
2. a. Large quantities of plant material accumulate.
   b. The plant remains partly decompose in an oxygen-poor swamp or bog to create a layer of peat.
   c. With shallow burial, the peat is changed to lignite (soft, brown coal).
   d. As burial continues and the load is increased, water and volatiles are pressed out, and the proportion of fixed carbon increases to form bituminous coal (harder, very brittle, black coal).
   e. Anthracite forms when the bituminous coal layers are squeezed and mildly heated (metamorphosed) by the folding and deformation associated with deeper burial and/or mountain building.

3. a. stratigraphic (pinch-out) trap; b. anticline; c. fault trap; d. salt dome
4. Oil in tar sands is much more viscous than oil in conventional reservoirs, so it cannot be recovered using conventional techniques.
5. Much coal contains some sulfur. When such coal is burned, the sulfur forms sulfur oxide gases, which undergo chemical reactions in the atmosphere to form sulfuric acid. The sulfuric acid makes precipitation very acidic.
6. a. directly (through south-facing windows)
   b. Solar cells are used to convert the sun's rays into electricity for home heating.
   c. Solar collectors (blackened boxes or tubes) contain fluids that are heated by the sun. The hot fluids are circulated through the home for heat.
7. It is produced by tapping naturally occurring steam and hot water located beneath the land surface in regions where the shallow subsurface temperatures are high because of volcanic activity.

REVIEW EXAM (Chapter 21)
1. The oil shales are heated (to about 480°C to 900°F) so that the shale oil seeps out of it.
2. A controlled fission reaction produces heat. Heat is used to make steam that turns large turbines. The turbines are attached to electrical generators so that when the turbines turn, electricity is produced.
3. Mechanical energy of falling water is used to turn turbines attached to electrical generators that produce electricity when the turbines turn.
4. Rising tides are allowed to enter a harbor and are trapped behind a dam. When the water is permitted to flow back through tunnels in the dam, its mechanical energy is used to turn large turbines attached to electrical generators, so that electricity is produced.
5. a. a potent source of heat
   b. large porous reservoirs with channels near the heat source and through which water can flow
   c. capping rocks that prevent loss of water and heat to the surface
6. building materials and industrial materials
7. nuclear, solar, wind, geothermal, hydrothermal, tidal energy
8. c
9. a
10. c
11. c
12. d
13. a
14. a. False. Many miners are still injured or killed each year; b. False. It is bitumen. Kerogen is the hydrocarbon material in oil shales; c. False. The primary fuel used in nuclear power plants is uranium-235; d. True; e. False; f. True; g. True

CHAPTER 22: PLANETARY GEOLOGY

VOCABULARY REVIEW (Chapter 22)
1. a. Jovian; b. terrestrial
2. nebular hypothesis
3. a. maria; b. highlands
4. lunar breccia
5. rays
6. lunar regolith
7. Cassini gap
8. Occultation
9. asteroids
10. a. comet; b. coma
11. a. meteoroid; b. meteor showers; c. meteorites; d. micrometeorites
12. a. stony meteorites; b. iron meteorites; c. stony-iron meteorites

APPLYING WHAT YOU HAVE LEARNED (Chapter 22)
1. It is any Jupiter-like planet with a very low density and a thick atmosphere.
2. It is an Earth-like planet with a very great density and a thin atmosphere.
3. It is a meteorite composed partly of silicate minerals and partly of iron.
4. It is any small solid particle in orbit within our solar system.
5. It is a meteoroid that has entered Earth's atmosphere at a great velocity and glows; a "shooting star."
6. They are planet-like bodies ranging from 1 to 100 km across.
7. It is loose rock fragments and dust on the surface of the moon, incorrectly referred to as "lunar soil."
8. A coma is the head of a comet.

ACTIVITIES AND PROBLEMS (Chapter 22)
1. gases (mostly hydrogen and helium, with melting points near absolute zero); rocks (mostly silicate minerals and iron with melting points exceeding 700°C; and ices (frozen ammonia, methane, carbon dioxide, and water)
2. Jupiter, Saturn, Uranus, Neptune
3. a. Area C is heavily cratered lunar highland; b. maria (each one is a mare); c. C-craters are less well-defined and do not have rays, unlike the B-craters, which are well-defined rayed craters; d. C (first), A (second), B (last);
    e. 1. As the earliest crust formed on the moon, it was heavily bombarded by meteoroids to produce densely cratered terrain, as at area C.
    2. Isolated impacts from asteroids fractured the newly formed and heavily cratered crust, so lava flowed to the surface and filled in the asteroid impact craters to form maria, such as at the A's. Higher portions of the early-formed, densely cratered terrain remained, as at area C.
    3. Upon final cooling of the moon's interior, a few remaining asteroid impacts shattered crustal rocks, sending rock fragments in all directions to produce the rays resembling splash marks. Crater B is such a rayed crater.
    4. Today, only small craters develop from meteoroid impacts.
4. About five billion years ago, a huge cloud of gases and rocky fragments began to contract under its own gravitational influence. As it contracted, the rotation of this nebular cloud caused it to assume a disk-like shape with a protosun in the center. Within the rotating disk, small eddy-like contractions formed the nuclei from which the planets would eventually develop. As the temperature began to drop, materials in the disk began to condense into small rocky and icy fragments. These, in turn, were swept up by the gravity of protoplanets and were incorporated into (accreted to) the protoplanets. Because of the higher temperatures in the inner solar system, the inner planets are made mostly of rocky materials and lack the gases and ices which are the main components of the outer planets.
5. Venus has a runaway greenhouse effect, caused by a thick atmosphere that is about 97% carbon dioxide.
6. Plants use up carbon dioxide during the process of photosynthesis, and some animals use up carbon dioxide to produce their carbonate shells and bones. Oxygen is produced by the abundant plants on Earth (as a by-product of the process of photosynthesis).
7. a. There does not appear to be any soil cover, only Martian regolith. There is also a noticeable lack of liquid water, but the winds are strong enough (from atmospheric convection) to transport sand and make sand dunes.
    b. The Martian landscape is different from a typical Earth landscape because it lacks obvious soil, plants, animals, and liquid water.
8. There would be meteor showers on Earth.

REVIEW EXAM (Chapter 22)
1. It is because Venus is similar to Earth in size, density, mass, and position in the solar system.
2. Mercury is not much bigger than Earth's moon. It is very dense, and its surface terrain resembles that of Earth's moon. Its surface temperatures range from as low as −173°C (−280°F) to as high as

427°C (800°F). These are the greatest surface-temperature extremes of any planet in our solar system.

3. Heat from Jupiter's interior and surface causes gases at the surface to rise upward and cool, forming the light-colored bands. The dark-colored bands form where air is descending and reheating.

4. Pluto is composed of a mixture of frozen gases and some rocky substances, so it is analogous to a dirty iceball.

5. the lunar seas (maria) and the lunar lands (highlands)

6. volcanoes, rift valleys, stream valleys, sand dunes, flat plateaus

7. a. Mars once had an active water cycle and active streams.
   b. Mars's subsurface ice melted and flowed to produce collapse features resembling streams.
   c. Mars's subsurface ice melted near volcanoes, the liquid water flowed to the Martian surface, and the surface water made the stream channels.

8. They are similar in size, they both appear pale green in color (from abundant methane in their atmospheres), and they have similar structure and composition.

9. a

10. d

11. a

12. a. False. It is not a Jovian or a terrestrial planet; b. True; c. False. Mass of the planet is also a major factor besides temperature; d. False. The "back side" is all cratered highlands; e. False. It is Jupiter.